10+1 Steps to Problem Solving

10+1

Steps to Problem Solving

AN ENGINEER'S GUIDE
From A Career in Operational
Technology and Control Systems

ANDREW SARIO

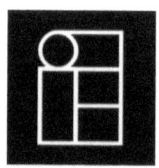

Engineering IRL

engineeringinreallife.com

Copyright © 2020 Andrew Sario

Andrew Sario asserts his moral right to be identified as the author
of this work

ISBN: 979-8-654-21850-6

Cover design by low-key

For Pharrah Jhae, Phoenix Elon, and Jarvis

In loving memory of my Mother, Erlina Villanueva Sario, 1954-2018, always believing in me.

Words to Think about when Reading this Book

"I like to talk about systems in saying that systems produce something which you cannot produce by just a set of elements by themselves. That means that there are system functions which are not the sum of other things and that the key then must be the relationships between the parts.

That sounds very straightforward now, but it has taken us a long time to get to something that fundamental about systems. Otherwise, people are talking about systems as any grand collection of things."

-Eberhardt Rechtin on Systems Architecture

"We are what we repeatedly do. Excellence, then, is not an act, but a habit."

-Aristotle

"I have been impressed with the urgency of doing. Knowing is not enough, we must apply. Being willing is not enough, we must do."

-Leonardo da Vinci

Contents

List of Illustrations

Unless otherwise noted, all images are in the public domain

10+1 Steps to Problem Solving

THE FRAMEWORK

10+1 Steps

WHY NOT JUST SAY 11? Are we doing math already? Honestly, I could have said 11, but the reason is that the first ten steps are the core process, and the "+1" is a type of anti-step. Consider the 11th a secret step. To gain the full value, you must only use it after completing the first ten steps.

Have you ever faced something challenging to deal with or to understand? That's a problem, literally.

People face different problems every day, so it is reasonable to expect they will become expert problem solvers through the natural course of their lives.

But I bet off the top of your head you could name several people who lack basic problem-solving skills.

How do you know that you're not the one who's incompetent if you've never worked on these skills?

Since it's a skill that helps both your engineering career and your life, it would make sense to improve it, right?

How can you improve your problem-solving skills?
With modern neuroscience, we know that the brain and the mind benefit immensely from experience. As we gain experience in our careers, we will inevitably learn.

But what is the structure of this learning process? And how can we actively take steps to improve it?

UNCONSCIOUS INCOMPETENCE	UNCONSCIOUS COMPETENCE
You are unaware of the skill and lack proficiency	Performing the skill becomes automatic
CONSCIOUS INCOMPETENCE	CONSCIOUS COMPETENCE
You are aware of the skill but are not yet proficient	You are able to use the skill, but only with effort

FIGURE 1 - THE FOUR STAGES OF LEARNING ANYTHING

It is possible that you have unconscious incompetence and the only way to shift towards conscious incompetence is awareness. Therefore, it's entirely possible that even if you believe you are skilled at something, you may have a gap.

So why not take an active approach towards improving it?

You could get training or learn by observing others. You can also read about it and practice alternative methods. That's where this book fits into your life.

I've been working on engineering problems for the past decade. Going through different phases of my career, I have found myself either directly solving, or being a part of the problem-solving process.

Aside from the big solutions I designed, the ones that came with awards, presentations, and patent applications, there were everyday problems to solve too.

After curating these scenarios and solutions, I soon realized there was a pattern. I became good at helping others find their way to the answers they needed faster. I was consistently asking the same types of questions to figure out what they should try or check next.

Thanks to this list of questions that I will share with you in this book, I was able to become the "go-to" person for solving difficult problems or be someone to bounce ideas off, hypothesize and strategize.

It started when I realized that only relying on senior engineers or the managers was not consistent, sometimes they were helpful if they had time - and that was a big if. Mostly their answers comprised of vague high-level responses that were probably correct but relatively meaningless to me.

Other times I would complete a task and find out afterwards, it wasn't what they were asking of me.

"Did I not ask the right question?" I would think to myself. How could I ask the person who just asked me to do something about how to do said task? I've spent five minutes pondering, and if I ask them now they might wonder what I was doing the last ten minutes!

I was always stuck thinking I know how to problem solve in general, but I don't know the specifics of this company, industry, or technology.

You could be a professional in industrial process simulations, but you likely are not experienced with every vendor software out there.

When I first started, after spending the first few days reading company documents, I finally got my first technical task.

"Hey Andrew, can you please connect this client to that server using this Ethernet cable? If it has issues, don't worry, I have some cross-over cables available."

I'm nodding along happy to get some hands-on work.

"And then can you RDP from the client into the server? Just log in to the local administrator account and use whatever IP's you want." He said.

I'm in my head at this point. I was a little anxious because I didn't know how to approach the problem.

Although I could build computers and had some lab work at university, at this moment, I could barely understand what half of those words meant!

I didn't even know where I could find power cables to turn the equipment on – was I allowed to plug into any power socket? What were the rules? Could I set up at my desk? Or was there a particular room?

You overcome these basic rules and knowhow as you work, but this type of story will continue for as long as you try to solve problems.

There's always a new technology or learning curve for the more experienced, so how can someone brand new even begin to approach this problem?

These are some pain points I was experiencing and are still relevant to this day. So why an engineer's guide/checklist for problem-solving?

What does an engineer do?

The number one thing an engineer does is problem solve. Now you may say, "Hey, that's the same as other professions!"

Admittedly, this would be true for virtually every single job on earth. As we already stated, all people are problem solvers.

Not to say other professions don't require problem-solving, they do. Still, when there are significant problems for a group or society to solve, using the latest technology of the era, historically, that is when you deploy engineers.

There are very methodical processes put in place and lessons learnt over the generations of problem-solving by engineers, including the Engineering Method that were developed by some smart people for this very reason.

In addition to the engineering methods, you will learn specific software, frameworks, standards and processes to solving problems from a company or technology point of view.

What a checklist like this will help you as an individual do, is direct your problem-solving and if you take the time to understand each of the steps in this book, you will be able to change both your mindset and how you approach problems such that:

a) you can reduce how long it takes you to solve a problem;
b) get better at solving the right problems; and
c) help your fellow engineers solve problems

As a bonus, you will also learn how to leverage these problem-solving skills into career opportunities.

Some problems require fundamental problem-solving techniques used in everyday life. Still, they can get more complicated. Maybe they involve others or a specific quirk of the system in a particular scenario.

You are going to face many problems as an engineer, and one thing you learn is not all problems are made equal.

Sometimes we have easy problems that are easy to solve, other times, its hard problems that are hard to solve. On occasion, you may get a hard problem that is easy to solve and finally, you may have to face the dreaded "easy problem" that is super hard to solve.

You can consider the *difficulty of a problem* as measured by the number of inputs or variables. And the *difficulty of a solution* as the time required to solve.

The Spectrum of Problems

We can derive four scenarios or problem types along this spectrum and is referred to in this book often:

Type 1. Easy problem / Easy solution
Type 2. Hard problem / Hard solution
Type 3. Hard problem / Easy solution
Type 4. Easy problem / Hard solution

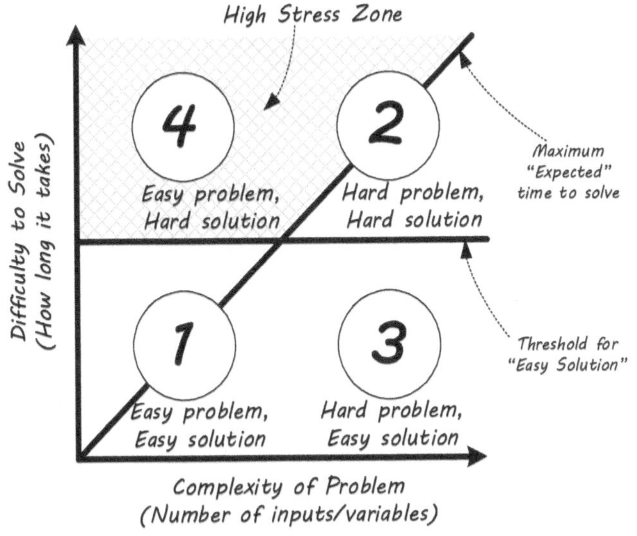

FIGURE 2 - THE SPECTRUM OF PROBLEMS CONCEPT

19

Unfortunately, in this business, if you've crossed paths with Murphy's Law, you know that the type 4 problem happens more often than you would expect had it not been for Mr Murphy.

The idea is to reduce the time it takes to come to the right solution, more type 1 and 3 problems instead of type 2 and 4. If it is a type 2 or 4 (hard solution) that will take a long time, you want to reduce that time and not feel like you are going insane repeating the same thing.

I created this 10-step checklist for this very reason.

How Grief Can Appear When Stuck on a Problem

Okay, right before I tell you the ten steps, I'll clarify that I didn't have ten steps in mind or some official textbook on this. I derived these steps as an observation of my experiences in the various stages you go through when stuck solving problems.

The exciting part is you could get a feel for the stage someone was in their problem-solving process, by how deep they were into the ten steps. There is a correlation between the "expected time to solve line" and the stage of the grief. The more of the list they have to use – the more desperate the person is.

Consider the "6 Stages of Grief" these stages are:
- Denial
- Anger
- Bargaining
- Depression
- Acceptance
- Finding Meaning

With time running down, people waiting for a resolution, you get asked several times questions like "what's wrong?", "why isn't it working yet?" "how long until you find the answer?"

You think to yourself – "if I knew the answer to any of these questions, I would have solved the problem already!"

Helping Someone Solve a Problem

When helping someone to fix a problem you may see them in one of the stages of grief, and I am using someone else here because it may be hard to identify distress in yourself at the time. It is a little dramatic, but it is the truth. This likely means they are dealing with a type 2 or type 4 problem scenario.

You know they are fixing something, and you know they are in one of the stages of grief; to help them, you should automatically get a recap of what they have done to approach the problem.

Meanwhile, you know this process and can categorize their actions into any one of these steps.

Now your answer to them is one of three options.

Option 1, offer them to try the next step on the checklist or one they haven't done yet from this list.

All of these are things one would find themselves doing typically.

It just might be that the one thing you point out is the unlock to solving the problem, or it could also be they just needed that brief reminder.

While they are dealing with the technical parts of the situation, they are hyper-focused on the micro details of the problem.

What you are doing here with 10+1 is offering them the macro part of the problem-solving process.

You probably don't have the time to babysit the actual problem-solving, but you've pointed them to the next step, and that's great. You are slightly more useful than before.

Option 2, offer them to repeat another step, particularly Step 1. This one can be hard to convince someone to do as the person has already done those steps. If they are in the denial or anger state, you may not get a friendly response. As it's often said, the sign of insanity is doing the same thing again and expecting a different result.

A true sentiment, but experience tells you that "doing it again" can fix the problem. The number of times I have personally helped someone solve a problem by just getting them to repeat what they've done is hilarious.

It's like I'm a magician.

Okay, this statement is far from the truth. Still, if it does work at this point, you can spend a ridiculous amount of time finding out why it worked, or you accept it worked and move on (most people do this).

In science, you need a few failures to have statistical significance, so it failing once may not be sufficient data.

Now you might think, "you didn't *do* anything". So to improve upon this, be sure you ask the person you're helping to repeat, but let's say it doesn't work anymore. (It failed, then worked, and then failed again).

Wait. Why did it fail again? Was it just a fluke?

Now we have more pieces of the equation to consider. "Was it a pattern, like on, off, on, off?"
"What was different between the time it did work and the times it didn't?'
"What if I did this a fourth time? Is this an anomaly or a pattern?"

This recognition is a crucial part of the process. You've set in motion precisely what Option 2 is— repeating a step. These new set of questions is Step 1 in the checklist.

To use this book effectively is to understand after completion of a Step, you can repeat a previous one. It is modular by design.

Now I know you're excited, but before you jump forward to Step 1 to begin your problem-solving journey, note whether you are helping someone solve a problem or if it is for yourself.

If it is for yourself, try to be aware of your grief. Avoid those six stages and focus on these 10+1 steps instead for better results.

The next challenge for you, if you are in the midst of a difficult problem is to take a breath, a step back and identify you are in a negative mindset in the first place, and reliably fall back on this checklist.

FISH vs. PDCA Problem-solving Techniques

There are a few problem-solving techniques that encompass the wide variety of general problem-solving methods out there.

F.I.S.H, for example, stands for "what am I asked to Find" "what Information do I have" "what Strategies and Skills do I have to solve?" "How reasonable is my solution."

You will also find variations of 4, 6, 7 or 10 steps to effective problem-solving forming some cycle of defining or identifying the problem, generating ideas, evaluating solutions, picking one to implement and then validating the result, generally along these lines.

The PDCA cycle stands for Plan, Do, Check, Act.

In other words, Identify problems, Test solutions, Study results and Implement the best solution.

Ultimately, we are all trying to do the same thing- find the root cause and fix the problem. The advantage of the 10+1 problem-solving method is it will get you to answers faster and more reliably.

And the best part is, it is compatible with other problem-solving techniques and enhances your effectiveness with them, too. It does not replace or conflict with other methods.

The Problem with Standard Problem-solving

These lists aren't wrong; the problem with them is they don't help in the bottle-necks of the process. It is slow and is generally too conceptually focused and detached from the reality of problem-solving.

For example, a typical list item is "come up with solutions and implement". That sounds nice, but it doesn't tell us which solutions we should try first.

Should we think of 2, 3 or 10 ideas? Do I make the implementation plan straight away? Or I pick one at random and repeat the cycle until we have a solution? This ambiguity leaves things to chance. It's too vague and doesn't help us solve problems reliably.

In the first step, I will give you the tools to ask better questions which will quickly lead to the next step. You don't list solutions yet.

This process will naturally cover all the ideas you would want in existing problem-solving lists mentioned, but with a bit more direction. The framework is one where steps can naturally flow on to the next, or return to a previous step.

How do I know when I can do each step?

Once you are familiar with the steps outlined in this book, you will initially go linearly, but as you become better at problem-solving, you can go to whatever lands you the answers the fastest.

Using the 10+1 method
You can go straight through 1-11, or do 1-4 and come back to 8, or 7, or 6. These steps give you so many permutations and combinations to try, even though

traditionally you might expect that once you complete a Step, you move on.

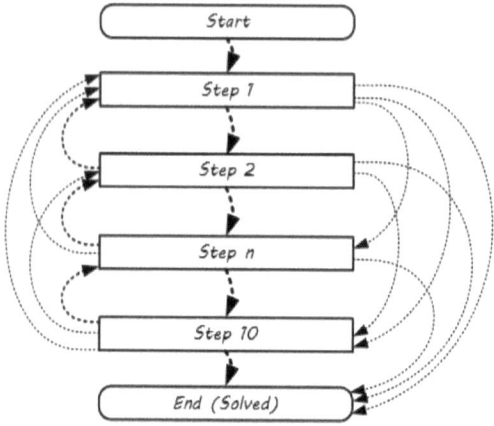

FIGURE 3 - 10+1 METHOD FRAMEWORK

Ideally, you would only go through all ten steps using the primary path, but as you work on trickier problems and more complex systems, this quickly increases the odds that you will need to repeat some of the steps.

It might be strange to think – if dealing with these problems can sometimes be so frustrating that it invokes stages of grief, why do it at all? Just imagine after days stuck on the same problem, the satisfaction when solving it, the release of tension.

It can be quite euphoric.
It might even possess you to say

"I...am....The greatest!"

Okay, I exaggerate, but after a while of repeating this narrative, you might actually believe it and if that's true – perhaps you can go on to solve more prominent and more complex problems or dare I say, *any* problem you face in life.

That is not a bad skill to have.

Oh, the third option? That's using the secret step.

Do you want to be useful in any problem-solving situation? Remembering this checklist is a good start, but reading this whole book will help you to be flexible and help solve any technical problem regardless of the industry, even if you are new.

This book will help you to be flexible and solve technical issues. No matter what industry you are in or how 'fresh' you feel to a role, this book will pave the way to a better approach to problem-solving.

Read each chapter to understand how to apply each step in a practical sense, what thinking goes into it and how to shift your mindset so you can embody the wisdom from the chapter.

After gaining this knowledge, to supplement your learning head to the Engineering IRL website and get access to free member resources at www.engineeringinreallife.com. The website includes the complete checksheet in various formats, which combines the checks at each step for you to use and some online challenges.

Now learn the steps and get insights on how to translate these problem-solving skills into expert engineering status.

STEP 1.

THE QUESTION

The Socratic Method for Engineers

Is the problem identified? - No, really, are you asking the right question?

Have you felt annoying for asking questions? How about: so irritating, your team called you out for it? Was a meeting put together to discuss your punishment? Were you sentenced to death because of it?

If you're reading this, I hope you haven't had the same fate as Socrates, because they condemned him for this very action.

Despite this being the case, we do want to be playing devil's advocate. Asking questions will help us get to the root cause of a problem, even if it might annoy others.

Are you still asking useless questions that make things complicated? That's a loaded question.

The power in this step and the overall problem-solving method is forming a downward slope tending towards the coveted answer.

This step is essential for you to understand because it is where all things start. It will most likely be the most returned to step out of this entire process.

After this chapter, you will have an understanding of framing questions better suited to solving problems. You will also understand after each problem-solving step in this book; you will return to this chapter and gain even more context to its power.

You will learn to ask better questions and ask questions of the very presuppositions allowing a situation to exist. These questions will serve as inputs to *any* problem-solving technique, not just the 10+1 method.

The Socratic Method uses the open type questions to form the dialogue so truths from an individual could come out, sparked by increased critical thinking.

The difference here is we are not exactly getting the truth of a person but the "truth" of a problem. It won't answer our open-ended questions on its own.

We can't apply the Socratic Method directly, but I will show you how engineers can use it to get results. First, we need to understand the types of questions that exist.

The Types of Questions

Questions come in different forms, and not all are equal. They serve other purposes and facilitate a direction of thinking, the line of questioning. What we want to do is stick to a line of questioning that point us in the right direction.

There is no official list on the types of questions; there are several lists you can lookup but here are the ones most common to you.

Try to think of a time you used one of these types of questions.

- **Closed Questions** – Engineers will call this the binary question. It invites one-word answers: Yes/No
  "Is it working?"
- **Open Questions** – This is the analogue question. You can't answer with a simple yes or no. There's more to be considered or elaborated.
  "Why is it behaving this way?"
- **Probing Questions** – This is the clarity question. You are encouraging someone to provide more context.
  "When did you notice this problem? Can I spend the next hour on it?"
- **Leading Questions** – These are magnetic questions that seem to push or pull the answer in a positive or negative route.
  "This is a huge problem, isn't it? Could this be a showstopper? We have a workaround, right?"
- **Loaded Questions** – I call this the trap card. It is a closed question but laced with an assumption. The Yes/No answer says yes to more than one thing.
  "Did you stop using hacky procedures? It wasn't a user error again, was it?"

- **Funnel Questions** – Start from a broad closed question to specific smaller questions to hone in on the situation. "Is it working?" (Broad) When did it happen? Did a change happen before this? Did you try the procedure?"
  Note: You can also use this to help the person breath, and take it slower.
- **Recall and Process Questions** – Memory questions. "What's the password? What do you normally press next? What's the process?"
  Note: You can understand someone's depth of knowledge.
- **Rhetorical Questions** – No answer necessary. "Isn't it nice when these things happen? I wonder if this is something obvious."

Understanding the types of questions creates a foundation for us getting better answers and also setting ourselves up to solve problems naturally.

Level 1 is spending time asking questions to whoever reported the problem, and then Level 2 is asking questions of the problem (to yourself).

Questions the Good Engineers Ask

So what types of questions do good engineers ask? A good engineer will use a combination of these question types to best suit the desired outcome.

If you've ever found you weren't sure what to do next, it's because you did not use the correct combination of questions.

You used one type, which doesn't move you any further than where you were before you knew there was a problem.

A broad closed question, for example, is quite useless on its own. "Is it working?" They answer "no".

If you didn't gather any more information, it might not have been worth asking.

The only time it's okay is when you have all the remaining data. If it's your 3rd time returning to this step, you tried a bunch of fixes and ask "does it work?" – This is when it's appropriate, and you shouldn't be asking any other type of question as it is wasting time.

Since it can be situational, this is a guideline for what to use in general. As you become more aware of the types of questions and put them into practice, you will have question-asking skills to envy.

Use in combination: Closed, Open
Use more: Probing, Funnel
Use less often: Leading, Loaded, Recall and Process
When should we use rhetorical questions?

As an exercise, think about situations where you felt you weren't asking the right question and remember the types listed and see what kind of question you could have asked to get what you were after, instead.

Next Level Questions

We just covered Level 1, triaging the situation, asking better questions of a person. For Level 2, you need to ask yourself questions to lead towards better answers and next step actions.

The purpose of the questions we ask ourselves is to question the presuppositions. After the first triage questions at Level 1 and clarifying the report, you now have a clear context to the problem, and you understand expectations of what they want you to solve and how soon.

Now you need to ask "*how do we know this is true?*"

If the report was the pump is broken, we know because we don't get the correct water level reading outcome. We ask ourselves:

"How do we know we didn't get the right water level reading?"

"Every time the pump is broken, do we always see this behaviour?"

"Is the reading the only indication of a broken pump?"

The presuppositions we are confirming here is:
- The water level is incorrect
- A broken pump leads to incorrect water level readings
- The pump is in fault according to other information sources (logs, alarms, visual observation or other correlating readings)

More on presuppositions and questioning inputs when we cover forming syllogisms in Step 9.

As a continuous improvement for yourself, be mindful of the types of questions you use, apply it to both the person reporting and to the problem itself, and you will begin to understand the power of better questions as it leads to better overall problem-solving.

The Legion of Fish

I was working at a Power Station with a cooling system supplied by a lake. Meaning instead of using a cooling tower or air-cooling with fans, they pulled in water from the lake, and use it to cool the process.

Then the water is returned to the lake after it had done its job, and was in a safe state to do so without negatively affecting the environment.

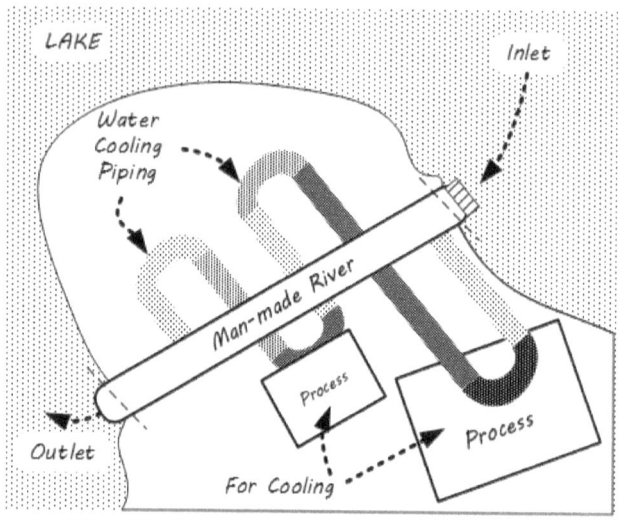

FIGURE 4 - WATER COOLING SUPPLIED BY LAKE

There was this human-made river going through the plant, end-to-end and as I peered in, I saw... fish.

More than I have ever seen, ever. More than an aquarium! If you had never fished before you could reach out your hand and be a master angler.

"How did so many fish get caught in here?" I thought to myself, and as I walked along this human-made river to where the water came in from the lake, there was a pinwheel type contraption with buckets pulling water in from the lake.

Each bucket, as expected, did have some fish. But not as much as I had thought.

Looking out to the lake there was a fishing boat nearby, but legally there was a minimum distance they must keep away from the plant.

Was this a coincidence? Why would the boat stay here? Wouldn't the fish distribution around the lake be the same? Since there was a minimum distance, they couldn't be getting fish straight from the inlet.

The truth was the water was ever so slightly warmer, which attracted more fish. So next I thought "is the water nearby so attractive leading to so many fish getting pulled into the plant?"

The number of fish pulled in from the river did not compare to the fish inside; the math didn't work out.

It turns out I had been asking the wrong question.

I looked at the point where the fish entered and focused my questions about why the entrance attracted more fish. I made a mistake. I assumed that the lake was the only source of fish into the plant.

Did you catch what the assumption was?

Excuse the pun, but what was the other source of fish? The staff weren't bringing fish from home and dumping them in, it turned out the source was... the fish!

Many of the fish stayed in the environment and reproduced. That is why there was so many more than the amount coming in from the lake.

STEP 1.
THE QUESTION

Let's focus back on this step. The Question. We know how to ask better questions now, what do we do exactly in this step?

Create a Narrative for the Problem

Creating a narrative isn't a long story-telling step or anything. You make a narrative saying, "here are my inputs, (triage by asking better questions), here are the questions that can be answered with one check and here are questions which depend on other information".

In other words, we ask a set of questions to form some inputs to your problem. You will have some unanswered questions that can either be answered with an obvious check or need to have other questions answered first. Each item tested has a possible result.

You don't need to go any further than this just yet. Simply create this narrative of inputs, questions, results.

The beginning, middle, end.

An example:

Inputs: Something doesn't work, and it happened today. No one made any changes before this. It isn't urgent, but it needs a resolution within a week.

Questions: Have we tried the three quick fixes I know? What happens when we do? Is the feedback appearing a certain way?

Expected Results: We expect the problem to resolve when we try this quick fix. If it's still not working, quick fix 2 is next, and we can check if it works again.

This combination of inputs, questions and results is the narrative. It shouldn't take you long to build this up in your head.

Don't Make a Plan

Don't make a plan (yet). This 10+1 step process is the guide and will have a little more planning built-in when it is appropriate. We simply build our narrative and get to work on the next step.

Now, if you are returning to this step, at some point, you need to reframe your line-of-questioning completely.

To Question or Not To Question?

When should you leave the current line of questioning? At some point, we need to change the questions and pivot, similar to the Legion of Fish story. Before you ponder and go too deep into considering every potential, do this only if you are returning here.

It can be tempting to try to reframe the question and spend time analyzing possible outcomes, but when there is a task at hand, and you've set up the narrative, it is time to get to work!

As a general rule, the time to question depends on the real goal of the situation. We established earlier that there are different drivers of the people involved with raising the problem.

Sometimes someone is biased and wants a fix to their problem using a specific product as a part of the solution.

Imagine someone tells you "My email software is failing to send". There are two problems to solve here.

First, the email software is failing to send, that's pretty obvious, and the second problem is the user cannot get a message to someone else.

So you have two problem-solving routes to take here – Fix the software problem, or find an alternative means to reach the other user with the message.

If they are particularly stressed, their goal might be to send that email ASAP.

Which solution do you pursue?
Both are my answer.

Here's how you would do it, you look at the technical problem and estimate in your mind how long it would take, while that happens you ask them to text or call the recipient with their message – depending on urgencies.

If it is an urgent message, but they need the email attachments, then you should send it via another client, even use your email address and then go from there.

If it wasn't an urgent email, but they want their software fixed because they have more emails to send, then this is where you concentrate. It all depends on the underlying goal.

How do you determine an underlying goal?

Use your open questions in a Socratic type manner. If you stick to the closed question, "Do you need this email software to work?" then they reply "yes". You end up focusing on fixing the software. Meanwhile, a deadline is missed, and you receive all the stress.

By using probing questions, you are letting them tell you their underlying problem. "What's the email you need to send? Who is the recipient? How urgent is the email? Does the email have to come from your address?"

If you get to their real goal, you can save their stress, solve their immediate problem and then solve their long-term problem, too.

Questions Subject Matter Experts Ask
A subject matter expert is someone who has in-depth knowledge in a specific area. As you solve more problems and gain experience in a subject matter, you slowly become a subject matter expert.

When there are problems in this area, you are the go-to person to answer the related questions.

Although this technique applies for problems you may not be an expert on, your subject matter expertise comes into play for the unanswered questions of the narrative you created in this step.

When you are new to a topic, you may only be able to come up with a couple of things you want to check. When you become experienced, you will typically have more.

Subject Matter Experts typically ask specific probing questions that eliminate potential fixes on the spot and is a big reason why they get to the answers faster than the average person.

You can't replace the experience of the subject matter experts, no matter how much reading or classes you take.

And to back this up, we now know in neuroscience there is particular brain circuitry that never lights up until you are in specific situations.

It means some experiences will enable parts of the brain which you would never access without the experience. Perspective is powerful.

The Soothing Sound of an Explosion

As a field engineer, you get to do long-distance travel alongside colleagues with different experiences than you. It becomes clear with the stories you exchange that you have different perspectives.

Some questions or assumptions you make are not what you'd expect because you lack the experience.

One colleague of mine fought in wars.

Fighting in a war is not an experience I've had (thankfully), but the conversation had sparked curiosity on my part. Just for team camaraderie, I asked questions about the stories my colleague was comfortable with sharing.

One such situation was the importance of digging trenches at each stop location. As they entered deeper into the enemy territory, it was increasingly important. We were discussing how it was "just another part of life" for him, the same way travelling to Site, as an engineer was just something you did or like teenagers going to high school.

The trenches were necessary because if grenades go off, the impact of the blast was more controlled. There were several other advantages and uses, but this was just one of them.

"Wasn't it scary when a Grenade exploded?" I asked. He chuckled, and with a big grin, replied,

"No, it was the best sound you could hear".

"You see, if you could hear the explosion, it meant you weren't dead".

Wow.

That made a lot of sense, but it was not the perspective I was even considering at the time. The inexperienced is reacting, while the experienced is proactive, knowing what situations mean and what to do next.

All this is to say that with your experience using this problem-solving method, along with your expertise in your specific subject matter, you will gain knowledge and become a go-to engineer. By being able to apply this method to various situations, you can quickly become dependable on more than one subject matter.

Deriving Questions from a Requirements Spec and keeping clients happy

I wanted to touch on this because not all problems you will face are going to be problems with a running solution. It may be a solution you need to create based on requirements, a commonplace activity in engineering practice.

Just because you have a requirements spec, it's not the whole answer. It still leaves many variations to consider. We are going to be asking questions to help form the picture.

To get better at deriving these questions, you will be using probing questions to the owner of the requirements. The questions will help you understand their underlying goals.

In this scenario, you use leading questions. You want to try to lead your client to the limit of particular requirements, not because you want them to go for it, but because you want to understand at what point they push back.

At what point do they feel pain?

You will get a sense of their real pain points.

Think about it from their perspective. There are goals they want to achieve, and they want someone to provide a solution which achieves those goals through a requirements specification.

The engineer focuses on delivering the requirements specification because it's in the contract, but it is one step removed from the underlying goals. This gap is how you under-deliver for a client.

Engineers love specificity and hate ambiguity. Therefore, it makes sense that they naturally focus on delivering the requirements spec, only.

You can run into situations where you feel like you over-delivered, but you have an unhappy client because you didn't meet their goal.

To get to your next level, you need to balance completing the requirements spec, as is your obligation, as well as meeting some of your clients' goals.

By understanding their underlying goals at the root of those requirements, you can also align your solutions to their goals, not just the minimum spec.

Part of your job is to try to minimize the gap between requirements and goals. Not just to make your client happy, but also because it will help you in the long run when executing the project implementation.

You will have less friction during the different project phases until you close out the project.

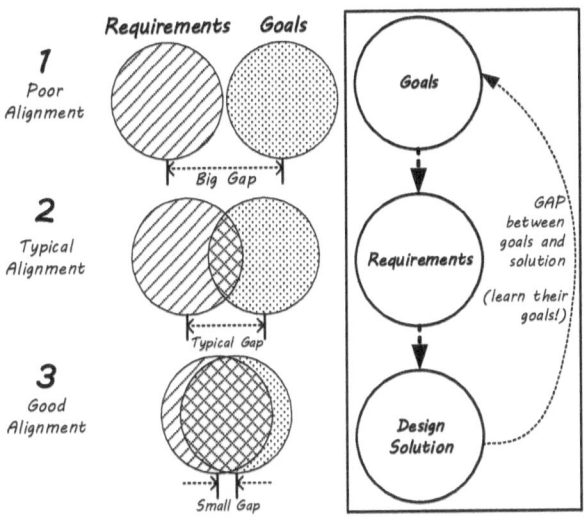

FIGURE 5 - REQUIREMENTS VS GOALS

These open discussions with your clients, while always delivering on the requirements spec, will help them adjust their expectations.

Either they change their goal slightly because their requirements make it unfeasible with the timeline or the budget. Or you favour solving a requirement in a particular manner even though it may not be best for another requirement, since one solution meets their goals better.

They will be happier. You will have some slack, and now you meet the requirements - with a happy client.

By using your new-found questioning skills, you can understand their underlying goals and drivers, have full clarity on the requirements spec, manage their expectations and put yourself in a position to know where you can over-deliver.

Questions When Inventing a New Solution

When inventing new solutions, the importance of this step skyrockets, this is because finding solutions is straightforward.

Usually, there are several solutions for different problems, and in fact, there are probably already several products or services available to address the issue.

In some cases, the problem statement is clear because it comes directly from the client. They could stipulate that they want to retain its intellectual property, so using another product or solution isn't viable.

Given this constraint, the keys to this step are similar to deriving questions from a requirements specification. You must understand the underlying goals of the stakeholders.

The stakeholders come in the form of the actual end-user and then the owners and maintainers as well.

As an engineer, you should repeat your questioning to each of these stakeholder groups independently.

"Independently" is vital here.

You want to isolate your engineers and understand their pain points.

Then you want to isolate their upper management or owners and understand their pain points.

Next, is the maintainers - get their perspective as well.

Finally, get the perspectives of the end-users.

These stakeholders are the users. You might even consider your company team as a user and understand how the business will support this solution long term.

What if the same person holds some of those roles?

Not a problem, within your conversation, ask

"As an *engineer*, what do you see as the most important aspect of the solution?"

"As the *maintainer*, what do you see as the most important aspect of the solution?"

If they are having difficulty separating and keep mixing perspectives, you can get them to detach themselves from the role by asking:

"If someone had to come in and take over temporarily while you were on leave, what do you see as the most important aspect of the solution?"

Now you have some actual feedback from the various perspectives. You can begin to see where things overlap and where they contradict.

Start with forming your problem statements around the overlap, as this will help you manage where the problems contradict.

For example, their user might say "I want the flexibility to make my own screens on the fly", and the maintainer might say, "I want to block changes without my approval, so things don't break!"

The easy way out is to give the problem right back to the client and say, "Hey, Mr Client, sort this out, is it the user option or the maintainer option?" Alternatively, I would find a solution around this obstacle, in addition to putting it back to the client.

Given the needs, we could make a section for user-changeable screens but are in a different area to the "official pages" that need to go through maintenance approval.

In short, you look for three solutions instead of 2. We can do user way, we can do maintainer way, or we can do this 3rd option, a compromise. Now go back to the client with these three options to help them answer and engage both stakeholder groups.

That was an example where someone else is the owner. What about for the engineer inventing something entirely from scratch?

The challenge is finding the right problem.

To find the right problem, you need to make sure you are asking the right question. How do we do this better? Make sure you remember this technique because it is also a great way to think about all of your questions.

The Physics Framework

You reduce your question down to unquestionable, fundamental truths, down to first principles and reason up from there.

It's similar to getting to an underlying goal of a user except since you are the owner and inventor, you are trying to look at real fundamental problems. An example of this is the advent of the automobile.

Imagine being Henry Ford in the early 1900s and the problem statement is "we want faster horses". Your solutions will be around horse care, food, training and breeding.

However, the fundamental question is "I want to get somewhere faster". It is the same with every single disrupted market; the problem statement is changed to address the underlying need instead, and then came up with solutions.

This change in the problem statement is how many multi-million dollar business ideas come to existence, with the power of asking the right question.

Now, as engineers the majority of the time you will be solving problems to existing solutions and not always inventing new ones. The whole 10+1 approach will give you the framework to solve problems along a product development process as well.

We. Vs. You. Vs. I

Last little pro tip here. I was purposeful in phrasing the question as "how do *we* know" as opposed to asking, "How do I know" or "how do you know" to someone solving a problem.

Affixing "you" or "I" puts direct blame on someone having his or her input information incorrect, triggering a standard defence mechanism that might quell the line of thought, allowing us to question.

It is as if you or they are undermined or did not do their homework.

By asking "we", it imposes "someone else" might have supplied the wrong information to us and so this is what using "we" will do. It detaches the blame. It also implies that we are in this together.

So, the next time you're working with someone to think through a problem, ask them, "but how do we know that is the case?" You will find this to be more beneficial for teamwork-based critical thinking.

The Power of the Question

Remember. The question is step 1. We must make a concerted effort to get this right and accept that we might not have the right questions in place when beginning an investigation.

Knowing this method of problem-solving, you will find as you complete every step, it is an opportunity to come back to step 1 and ask better questions.

If you haven't yet solved a problem, but you have more questions, this is a good thing. It is better to be finding a solution where something is wrong than trying to find a solution to a problem you can't find.

Do not forget someone's account of the situation is riddled with biases or concerns they have. They might not think something is a big deal and can miss a detail or they can focus on the part of the problem requiring action from them as opposed to someone else.

STEP 1.
THE QUESTION

Your checklist for this step

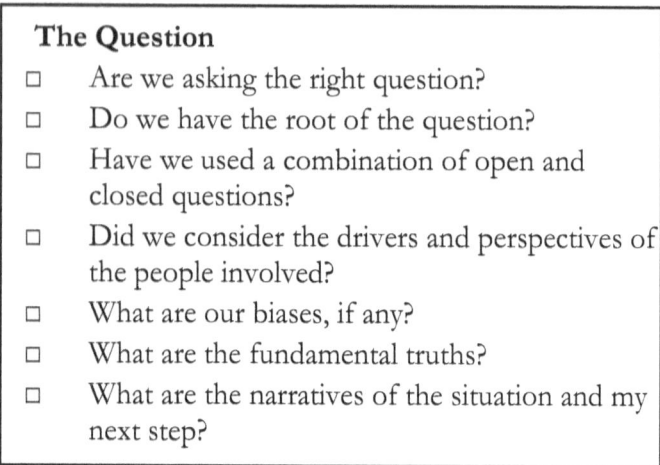

The Question
- ☐ Are we asking the right question?
- ☐ Do we have the root of the question?
- ☐ Have we used a combination of open and closed questions?
- ☐ Did we consider the drivers and perspectives of the people involved?
- ☐ What are our biases, if any?
- ☐ What are the fundamental truths?
- ☐ What are the narratives of the situation and my next step?

It may seem a lot to think about, but this entire question process will eventually only take a few minutes. Rather than analyzing the situation for hours, struggling, you will develop questions in a few minutes, setting in motion the steps to finding a solution.

After establishing some vital questions, it is time to investigate and test possible solutions. We don't dwell and randomly select the "best solution" as this is unnecessarily wasted time. So where do we start?

We start with Step 2—the Obvious.

STEP 2.
THE OBVIOUS

What Should Work?

Have you applied the known, related troubleshooting step to solve the problem?

That does sound silly, but it is possible to start thinking about 50 items up the chain of things that could be wrong. Our confirmation bias tends to look for data already supporting our hypothesis despite the data showing us otherwise.

In this chapter, we learn not to miss the obvious and waste time investigating the information. Think about those times when you weighed up several options you can't know the answer to until you try.

After several minutes of pondering, the first one worked anyway!

THE OBVIOUS

We force ourselves not to look around yet and try the obvious. This step will save time looking at the wrong problem.

An example is; you set something up, something goes wrong so you look for other things to be the fault because it can't be what you just set up, even though the data suggests otherwise. The data, in this case, are the symptoms of the problem.

The funny thing is, as you grow in experience, this becomes more of an issue - more on that later.

The Logical Slippery Slope Fallacy

Imagine you are working on setting up a piece of equipment, and the order is something like:

1. Wire everything up and power on
2. Run a configurator
3. Run the device

After running the device at step three, we get an error "configuration missing, please run configurator".

You might think, "well I just ran the configurator, so the problem has to be somewhere else".

So you go on a witch hunt – maybe there's a firmware version, a different configurator, perhaps you have to rerun the device, but it didn't load properly, etc.

But let's take a quick moment to forget our confirmation bias. What data do we have?

An error message displaying "configuration missing, please run configurator".

I know we already ran it, but given this data, the only conclusion is to run the configurator. It is so obvious it can't be the answer. Nevertheless, this is where you should start.

Perhaps it still fails, and that's okay, we can move onto the next step. However, it is a fool's errand to investigate beyond this point before double-checking the obvious.

The reason why more experienced engineers are susceptible to skipping the obvious is that they are more experienced.

They have seen this situation before; they know if you do steps 1, 2, 3 it works, they have seen it hundreds of times. And if you don't follow 1, 2, 3 it doesn't work. So the fact they followed steps 1, 2, 3 they conclude it must be something else.

That's solid logic.

Remember though; logic is only as true as its inputs. So whilst it is likely the case, check the obvious.

Another example is when setting something up or configuring a product, there is what we call known problems.

Of course, as things change bugs get resolved in future versions or designs of a product, but it costs money to upgrade. Therefore, an older version may have a known problem – with a known solution.

Before digging any further, make sure you have gone through all the known solutions. Sometimes there are "known" or standard troubleshooting steps.

STEP 2.

THE OBVIOUS

Suppose you were typing on your keyboard and pressed the wrong key. The obvious step is to press backspace and type it again.

OR

You could assume there's a mechanical issue with your keyboard, or there's an issue in the driver, or even worse, somebody hacked you!

An extreme example and sounds ridiculous, sure. Still, I challenge you to take any problem you have and consider if you have missed any obvious things and are mulling over much bigger and more complicated answers than need be.

The Serial Obstacle

I worked with someone in a lab environment once on a project requiring checking serial numbers and then placing a label at the front of the device with the number.

The majority of the equipment was in a cubicle with the back close to the wall—about the 1-standing-human-sized gap.

There were a few boxes with equipment sitting on top next to the cubicle, and the front had full access.

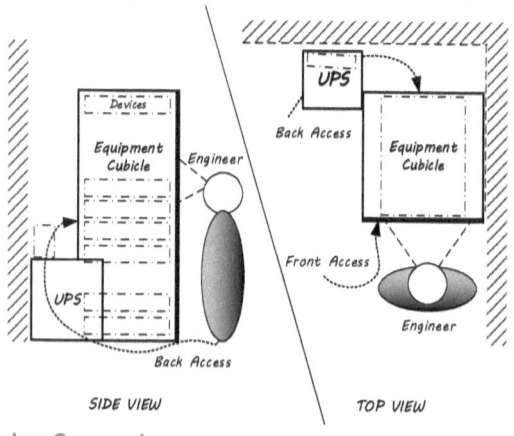

SIDE VIEW   TOP VIEW

FIGURE 6 - LAB CABINET LAYOUT

When standing in front of the cubicle, information on top of the devices were easily observable by the engineer.

I estimated this was a 15-minute task, and he was well on his way to completing the job, working from the top of the cubicle down.

Read serial number, label.
Read serial number, label.

One after another, the devices are labelled. I leave at this point expecting to come back, and all devices are labelled.

I'm interrupted with a small task, so after an hour I come back to check, feeling bad because the last forty-five minutes he was probably twiddling his thumbs waiting for me.

"How's the labelling going?" I ask.

"Pretty good, I've got about 90% of them done." He replies.

"Great …wait, what?"

He goes on to explain what the roadblock was. The issue was the serial numbers were not visible from the front for the last couple of devices towards the bottom of the cabinet. They were on the back.

If you recall I mentioned there were some boxes beside the cabinet with equipment on top, preventing access to see the serial numbers on the back of the devices.

The thought process was, "I can't access the back, but there must be some special way we do things". There are boxes in the way, but the equipment on top might be heavy. Is there a "slide-out" feature of the cabinet?

Now applying this step, you might look at the data in front of you, which is "I need to go from where I am to this next position, but there is a box in the way". The obvious step here is to move the box. Yes.

Here's where I encourage trying things. We are trying the obvious. Most times, we look for some big official plan, implementation strategy or an example of the task done a certain way, so we don't make a mistake.

More often than not, you can solve a problem right here at this step by trying things and see what sticks.

Try the obvious things.

Let's look at some of the box moving concerns – "it might be too heavy" we can check this by trying to move the box. I am in no way suggesting to rush things unsafely. I say this mainly for the things that are quick and safe to do.

Solving Problems Using Only the First 2 Steps

By combining Step 1 and 2 of the 10+1 problem-solving process, you can solve many problems that should be a type 1 easy problem, easy solution and not anything more difficult (remembering the spectrum of problems in the introduction chapter).

By taking more time than necessary, you risk increasing your time to solve and crossing the expected time to solve line on the spectrum of problems.

Here is an alternative approach using these steps.

Assume the box was indeed too heavy. The assumption baked into the question is "I need to stand at the back of the cabinet so I can see the serial number".

Logically you can't stand there, so you have no solution, but let's look at the question again, step 1, "I need to see the serial number of the device".

We removed "standing" in position from our question, so now step two; what is an obvious answer here?

"I need to see something where I am not physically standing".

Is there any type of device or technology to assist with seeing something? An image perhaps?

Aha, I have a device in my pocket with a camera; I can reach around and take a photo. Then I can see the serial number in the picture—no need for standing at the back.

Now before you stop right there and say "but what about if my arm can't reach? What if it's too dark?"

Just try. Try the obvious.

You'll have answers shortly and if these problems arise, deal with them next.

Common Sense Isn't Very Common

In engineering, it can feel like solutions are so trivial. Indeed, common sense would prevail here, right?

A sign next to a visible puddle of water stating "slippery when wet" is essential to make it more visible, this is merely an administrative control, but it is a necessary one nonetheless.

The control is necessary because there is a risk, and appropriate mitigation needs to be in place. At some point though, someone might trip over the "slippery when wet" sign.

Do we need another sign for this sign?

That would be ridiculous; surely, common sense comes into play here, right?

Engineers must consider these "common sense" scenarios when doing risk assessments and select appropriate controls from the engineering hierarchy of controls. We must define the levels of risk, have the responsibility of choosing the proper controls and essentially determine the value of human life.

Truly a huge responsibility. The point is, we don't assume common sense exists and we don't assume the obvious checks won't yield results.

The Donkey Tunnel Conundrum

The obvious does not mean a shortcut. We are merely reserving further analysis until completion of the obvious checks.

There's a story about users of a tunnel who have donkey powered carts. The means of transport is by donkey. They commute through a tunnel, but there is a problem. The tunnel is a little too low, and the donkey's ears get hurt, scraping the roof.

Several users complain about this, confirming this is a real problem. The tunnel owner must fix this as he charges a toll fee.

He then applies the "obvious" fix.

A strategy is marked out, and he cuts out two slits along the roof of the tunnel. Enough space for the donkey ears not to scrape the roof as they walked along.

The next day as the owner proudly looks at the updated solution, a tunnel goer asks, "What are those two cutouts for?"

"Those are for the donkey ears, obviously."

"Will this work for my horse I have scheduled for later?"

Obvious might not be so obvious.

What is an Obvious Task?

The obvious fixes referred to in this step are the *known* fixes. We are not inventing new obvious solutions (which is highly dependent on asking the right question).

For the donkey story – the question answered was the donkeys' ears scraping on the roof.

The better question is around the actual problem; the height of the tunnel is not suitable for different animals, which leads to either solving the tunnel clearance or shortening the donkeys' height (remove the shoes?).

The point is, we aren't inventing a new solution. You are going with the tried and true solutions here.

For example, if you hear a tap leaking, everyone knows you should tighten the handle first.

You don't go digging up the ground, checking the pipework, go outside and check your inlets or check other taps before tightening the main ones.

THE OBVIOUS

We are avoiding wasting time pondering down the chain of problems.

You might think, well there's only 1 or 2 obvious checks for this given situation, and that's fine, do those 1 or 2 tasks first before going any further. These are the main tasks, the known solutions.

Is this always going to be a one-check step?

How the Obvious Expands with Experience

The obvious is a list of your known solutions for a given situation, system or problem. Naturally, as you work with these systems, you gain familiarity and experience. With this experience, if you learn more known solutions from either your experiments or working with other engineers, you add them to your "list of known fixes".

I encourage you to ask your fellow engineers when working with them, "What are the obvious things they would try first in this scenario?" It doesn't matter if you already know a few things and it also doesn't matter if they are junior to you.

Different life experiences mean different perspectives, and you prevent yourself from learning alternative solutions by not seeking them out.

Build your list of obvious known solutions as you gain experience.

How to Turn Experience into Expertise

You continue to gain experience and build your list of obvious fixes. You pretty much know all the known solutions applicable to a given situation, system or problem. How do we measure expertise or known when we've become an expert?

To be considered an expert isn't just a function of time. It may be indicative, but not a reliable measure. An expert is someone who has a depth of knowledge in a specific area.

Someone with a wide range of knowledge on a topic but without depth couldn't be considered an expert.

To give you a model to work with there is the P-I-F-T method which stands for:

- Processing of relevant information
- Interaction with relevant information sources
- Feedback with relevant metrics
- Time spent on the above

You assess your level of expertise against this list by addressing all the above and comparing with someone you consider an expert.

A non-expert could be a skilled or experienced engineer. The distinction is the level at which they operate or perform.

How an Expert Engineer Processes Information

You can consider one an expert if they perform mental operations beyond what non-experts would do on a piece of relevant data.

If presented with a problem with a particular application, a non-expert might assess there is a problem. An expert will determine this too but also consider possible solutions, risks on related systems to the solutions and think of what would happen next.

This assessment may be the more technical, calculated or skilled portion of an engineer's expertise.

How an Expert Engineer Interacts with Information

Presented with the same situation, let's say there is an inspection happening on a process plant, a non-expert may only notice something obvious like where the primary device is.

In contrast, an expert is looking at minute details, seeing things about the environment, related equipment, what processes are also around, model numbers, behaviours of systems making up these types of equipment.

An engineer observes details and their experience dealing with similar situations show nuances of the situation they are dealing with based on previous experience.

How an Expert Engineer uses Feedback to Make Incremental Improvements

A feedback loop is a critical component in any control system. You need a way to know where you are so you can course correct.

A non-expert might make mental notes on doing things one way or another based on what they have experienced in the past. They may know how to get jobs done and have had lots of success solving problems.

An expert is one who seeks out lessons learnt, looking at their experience as well as others to see if there is new information from recent projects so they can adjust.

More often using a feedback loop to know they are still solving problems in the best way. They devise metrics for themselves to know if their solutions are still the most appropriate.

Developing Expertise over Time

As you can see, expertise does not only develop over time on its own. It develops over time with the conditions above in P-I-F. For you to build your expertise, you can't just rely on the experience of 12 years, although this may gain some expertise you can either accelerate your time to get there or be in a better place in 12 years compared with if you go with the flow.

If this is true, while we execute Step 2, we take a mental note and then a physical reminder to position ourselves towards expertise.

Why am I explaining expertise?

The first is we are trying to gain mastery of our problem-solving, and an expert will problem solve a specific field the best.

The second is because this book is not just a process of problem-solving; it is bits of gold and advice for you to continue your development towards an expert problem solver.

The first time you read these steps, you might be thinking tactically, which is perfectly fine as it is a massive advantage of 10+1.

STEP 2.
THE OBVIOUS

After actively solving problems with the lessons in this book in mind, you will notice different things reading the second time through, similar to when an expert interacts with the same information they see details the non-experts do not.

But just because someone is an expert, it doesn't exempt them from missing things. There is one problem with becoming an expert when dealing with Step 2.

The Expert Trap

If you are an expert in a particular field, you are by definition, the best person for the task when it comes to solving a problem in that area. The trap is when the status leads to the ego, and you skip over the obvious things or make assumptions.

You may struggle identifying areas you are not an expert in and spend too much time not reaching out since you are considered an expert in other areas by your company. The main thing is to become aware of this, so you have a chance of recognizing when you are in one of these situations.

Becoming an expert problem solver requires understanding how an expert deals with problems, emulating and embodying these principles even when it comes to the obvious steps in problem-solving.

Getting Paid the Big Bucks

The expert engineers tend to get paid the big bucks for both finding solutions to complicated problems and wearing the responsibility of a solution.

However, there is a simplified definition of an expert, and it might be the most measurable way to think about it.

P-I-F-T gave us a framework for what an expert may look like across a few markers, but one definition by Alterman is:

"Somebody who obtains results that are vastly superior to those obtained by the majority of the population (of engineers)".

The more experience you gain, the more items you can add to your list of "obvious fixes", and the more you can align with the skills of an expert, the faster and more reliably you solve problems.

Your list is how we correlate your problem-solving skills to becoming the "go-to" engineer and subsequently, an expert without sitting idly for twelve years. You bring value and become an asset worth paying the big bucks—more on this in the "Engineering In Real Life" chapter.

The Importance of the Obvious

Your ability to fix things in Step 2, will reduce your average time to a solution below the "expected time to solve" line on the Spectrum of Problems outlined in the introduction.

The reason why this process enables you to solve problems faster without missing steps is that the process encourages appropriate actions at each stage.

When starting, you may only be in Step 2 for a short while and typically move on, but as you gain experience and expertise, you will solve problems right here, without wasting time.

STEP 2.
THE OBVIOUS

Your Checklist for this step

The Obvious

☐   Have we applied the known steps to resolve?

☐   Did we try doing exactly what the clues say?

☐   Did we follow our basic procedures in detail?

☐   Did we try things to see what sticks?

☐   What would an expert check?

☐   Is my ego or pride preventing me from performing a quick check?

☐   Take note of the "obvious" checks and develop your list of known fixes.

Thinking like an expert and building your list of obvious checks will let you solve the majority of problems, right here at Step 2.

The more problems you successfully resolve, the more success itself is associated with you. You can go on to enjoy solving problems at the obvious, as an expert on the matter.

If the problem remains, you either return to step one since you have new information from this step, or you move on to the next step.

The obvious checks looked for the known solutions, where you can see. It's time to go into the unknown where you can't see - it's time to get some eyes on the situation in Step 3.

STEP 3.
EYES

Eye of Horus
You can't fix what you can't see.

Have you set up the correct logging, sensors, indications, report triggers and alarms so you can see what's happening?

Have you checked all the related moving parts of your equation? Do they have any way to be monitored? What's the actual status?

Now that you have done all the known basic troubleshooting and still have a problem, this is an excellent time to get some eyes on the situation. To confirm your assumptions while it's still early.

This step is vital because it will help reduce your time to solve due to flawed assumptions. You can find yourself chasing a red herring because you might be following solid logic, only to find out the whole time there was a completely different starting situation than initially thought. You check your inputs early because they drive the rest of your problem-solving process.

It is unreasonable to work blind after doing the first two steps. This step has enough value; you can already return to Steps 1 and 2 afterwards.

Suppose you are still having problems after you've built up your logic and are confident it should work with your actions so far.

In that case, you could have an illusory correlation bias - you inaccurately perceive relationships between unrelated events. You can avoid this by getting your eyes on the situation.

How to get Eyes on the situation

For civil, mechanical, electrical, and aeronautical engineers, this means having the right sensors in place, tools for measurement, making useful observations which you can only get from the field.

For systems, networks, telecommunications, cybersecurity, IT, controls and automation engineers, you need your logs from devices, network traffic captures, reports, alarms and system status.

Most importantly, for all the engineering spaces and general troubleshooting, you need to be able to reproduce the problem.

If you can't reproduce the problem, don't worry, we will get to that but, this is still a necessary step.

If you can reproduce the problem, you need to run over your assumptions, reproduce the problem and make sure all results align with your assumptions.

Press where it Hurts

If you believe you have a flow rate problem due to a faulty pump, then your pressure sensors at both the inlet and the outlets should behave per your predictions.

The behaviours should also align with any error messages or logs you can get from a monitoring device.

If you can reproduce the problem and you can get the right eyes on the situation, you will be able to spot the outlier from nominal behaviour and dig from there.

"Press where it hurts".

Think of a doctor diagnosing your ailments, and he presses to find where it hurts. Without squeezing, you are relying solely on someone's report of their symptoms.

Dealing With a Problem You Can't Reproduce

So what about the other case? We have a sporadic problem which we can't readily reproduce. It happens a few times a day or once every blue moon.

Well, this becomes the most critical step.

The last time the event occurred, it may be hard to know what happened, unless by chance you were looking at the right place, at the right time and saw it.

We want to avoid relying on chance. The whole point is to decrease the amount of time we spend solving the problem.

You need the appropriate monitoring at the point in time the problem occurs. If you have this setup, it means when it happens again, we have a means of recording the situation for post-analysis.

Vision is so important, and you know it to be accurate because the first thing you do when you enter a dark room to find something is to turn the lights on.

The other ways you may find something in the dark is to use your ears if it's your pet.

Or maybe you use your sense of touch and feel around for the object.

You can even combine other senses, including smell, to find something faster.

All that is to say, seeing something can also mean a way of identifying an anomaly. Don't confuse this step with only the visible spectrum of light.

At this point, you may be wondering, "Does this mean we just put in the sensors and all we can do is wait?"

For problems you can't reproduce that happen spuriously, you may not have the luxury of time on your side.

Or the consequences of it happening again without a plan are unacceptable.

If that is the case, then these are the only two things you can do:

1. Ensure that if the problem were to happen again, that we mitigate or remove the impact.
2. Use a test environment with as many of the variables possible simulated or duplicated.

You may have to rapidly accelerate your problem solving to Step 7 or jump straight to it. There you will know how to approach reproducing the problem.

Sleeping on a Missile

I was in a small town somewhere between South Australia, the Northern Territory and Queensland, where I worked with a couple of engineers to solve a mysterious network problem.

The problem only happened every so often and was last reported to us about two weeks prior.

It took a week of remote diagnosis and then another week to mobilize all parties to be on-site at the same time.

I attended as the role of vendor networking expert, and the other two were consultant-type owner's engineers. This relatively complex system was live and providing power to the town and a nearby mine.

This dependency meant like in many critical infrastructure projects, we couldn't easily shut things down and had to rely on more passive means of investigation.

We already completed Steps 1 and 2 of this book extensively with the remote diagnosis and troubleshooting weeks prior, so what have we learned is the next area of focus?

Step 3: Eyes.

I wanted to get more eyes on the situation, and my contemporaries were keen to analyze this data to figure out what the problem was.

We would develop some data visualizers with Python and run comparisons of the data.

We had a plan, and we placed what we will refer to as "network taps" in some critical locations. We knew we were getting timeout errors every so often and had to find the root cause.

We were looking at the data for any strange behaviour. After a few hours of data collection and the error occurring twice while we were there, it was time to hunt! We combed through the data and…

We found nothing.

Nothing strange, it all looked like nominal behaviour, even at the point of error. We thought we might find a spike of traffic or some error message, but nothing.

So we moved on from the network at this point and looked at the system reporting the error, maybe it was wrong?

We spent the entire day following this path and still had zilch. We did, however, have at least five more things to try down this path the next day.

At dinner, we discussed the rooms we were staying in and compared the quality of our accommodation. For context, it wasn't great, but for these two engineers, their last project was six weeks in a submarine and on some long work stretches, found themselves taking a nap on a missile.

"Having a bed is more comfortable than sleeping on a missile!" they said.

"Was that the hardest part? Uncomfortable sleep?" I asked.

"Nah, feeling disconnected from the real world is harder."

One of the challenging parts was the fact they could only have one fifteen-minute phone call per day. The limited window was because it had to be when they surfaced.

Combining the limited amount of time the submarine could spend on the surface and all the crew members on board, time slots were limited.

This limitation meant only so much data bandwidth for communicating with their families. Your conversations were limited to priority messages, only.

With such small amounts of data compared to the entirety of the situation, the engineers had to assume the rest of the information they received back from their family, based on the priority – or critical data.

Instead of a conversation about how their day was in detail, it would just be the highlights.

STEP 3.
EYES

So what does knowing this have to do with our problem? Well, maybe we were only dealing with a small sample size of the data and didn't have the full story.

Whilst we did place our network taps in the critical locations, I figured we should go back and put some extra network taps in some of the less critical locations.

The odds of them carrying a clue when they weren't part of the core communication path was low, but it was well worth expanding our search, increasing our senses.

We were only looking at the priority messages.

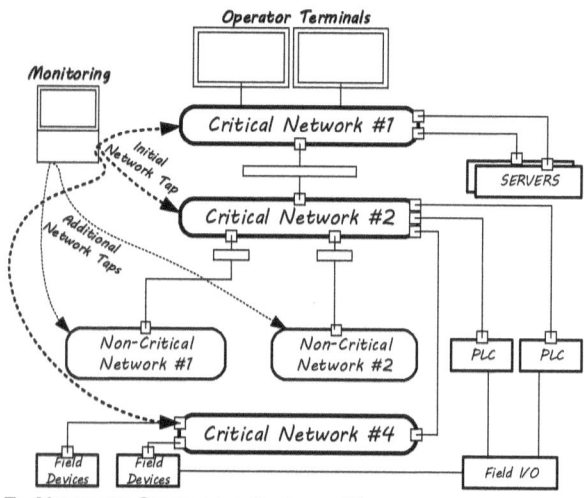

FIGURE 7 - NETWORK SETUP AND PRIORITY MESSAGES

Based on the logs collected initially, we assumed it was not network related.

This assumption is why we spent a day searching elsewhere, but can we confirm our assumptions?

The next morning we had a choice, continue from yesterday doing the five things we planned, or go back a step and investigate again.

We chose to investigate and get more eyes on the situation. If for no other reason than to prevent the idea nagging us in the back of our minds.

If nothing comes up, we can pick up where we left off.

When we got the data from a few additional non-critical locations, we began the hunt once again.

Still, before we could run it through the same level of data analysis, we found something peculiar right away.

It was a packet from a protocol that should not exist in our network.

It turned out one of the less critical networks had a device installed at a different time than the rest.

It was configured 99% correct but had one feature left turned on. This feature is forbidden in this type of network because it has the potential to cause issues.

We found the problem, and we didn't waste any more time investigating the wrong path.

Even if we didn't solve the problem right here, we had a strong basis for moving onto the next steps.

If you have the ability or opportunity to get more sensors or data collection in place, don't underestimate the value. Just collect it.

Get more eyes on the situation.

How to Know Where to Place Your Eyes

As we just covered in the previous story, we placed eyes on the situation, but we needed to cover a particular spot to reveal the truth.

So how do we know where we should place our eyes? After all, if you already knew you would have that area covered already, right?

There are two techniques you can use at this step to get better at placing your vision.

Way number one - zoom out.

Consider a topological view where you are looking down at all the pieces. You zoom out to a point where all of the parts are visible. You lose detail, sure, but you see a bigger picture. The purpose of the eyes is to illuminate this detail, so you don't have to worry.

You think about it from a strategic standpoint.

Make a high-level sketch of your situation, zoom out of your scenario; look at all the conceptual entries and exits, where energy changes forms or where data or processes change. It will become obvious where to set your data collection.

Way number two – change your perspective.

I don't mean top view, side view, etc. I mean, imagine you are the process itself. Put yourself in the shoes of the process. This process is the same way Albert Einstein changed his perspective to someone moving at the speed of light.

You will also perform a thought experiment. But this time from the perspective of the problem itself, propagating through the use case.

You imagine all the points of interest from this perspective and use it as a guide to determine if you need to place some additional monitoring at this spot.

How Easy is it to Turn on a Light Bulb?

Someone once asked me "how hard is it to turn on a light bulb?" in an attempt to describe a simple process.

"I don't know why he couldn't figure out where the problem was, you only flick the switch and the light comes on!"

"How easy is it to turn on a light bulb? The problem is either he didn't flick the switch, or the light bulb is dead!" he said.

He was stating the simple process of turning on the light meant there were only 2 points to look for a problem.

Either an issue flicking the switch or a fault with the bulb.

But it is a little more complicated than that.

Take a look at the electrical distribution diagram and note that you may have seen some of these devices or structures around when driving.

All of these components are geographically placed kilometres away from each other, with the power generation plants hundreds of kilometres away.

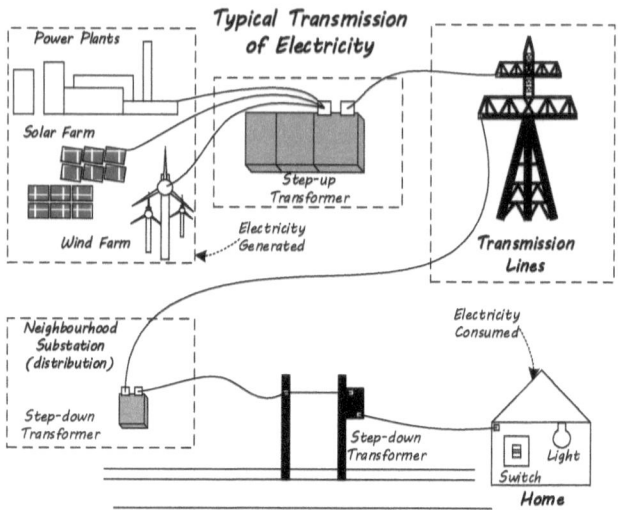

FIGURE 8 - POWER GENERATION ELECTRICAL DISTRIBUTION

If you've got a new bulb, flicked the switch on and the bulb still doesn't light up, you might have a problem with a fuse in your house.

If that's fine, it could be a problem in a transformer.

If that's fine, it could be in the transmission lines themselves.

If not the transmission lines, it could be at the distribution lines.

If it's not at the distribution lines, it could be at the power station not generating electricity.

If it isn't generating electricity, there are several other problems it could be in there.

How your home gets electricity can be a difficult concept to explain. Every bit of electricity you use is generated at some power station.

In its basic form, imagine someone riding a bike with a generator attached which produces electricity as they pedal.

Consider this scenario - every time you want to turn on your light. You call a friend to start pedalling on the generator bike.

They have to keep pedalling for the duration you want your light to stay on.
It's not connected to a pre-charged battery. (Things are changing with technology on this front, but we use this simplified scenario for the analogy).
As they slowly tire, your bulb starts to dim, and eventually, they stop and your light goes out.

Fortunately, we have the technology to produce electricity more efficiently than people on bikes, but either way here's the rest of the example.

If the bulb doesn't turn on when you flick the switch, the average person might say, "If it's not at the house, then the problem is with my energy company" and leave it there.

Think about the actual size of that "electrical company scope" in the next diagram, both geographically and the amount of energy and engineering in the system.

We often look at a scope as our direct vicinity or what is right in front of us. Our brains categorize things quickly, and anything outside of that definition is deemed irrelevant.

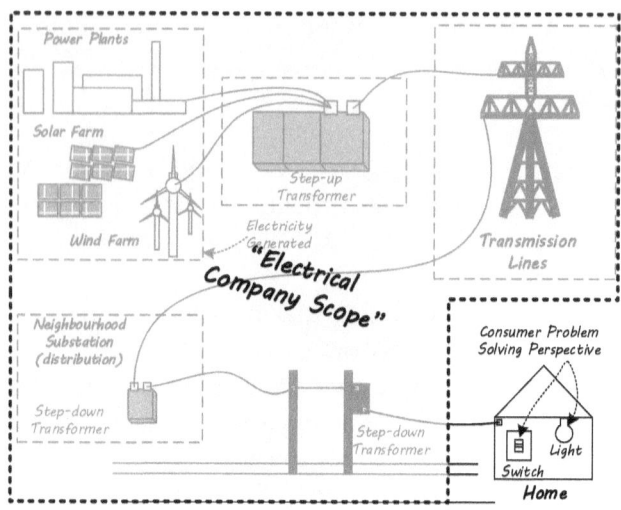

FIGURE 9 - USER PERSPECTIVE ON POWER

But in this scenario, you are the engineer finding the root cause. You are an energy company. You would want data from each of those points of exchange noted in the example.

Do a thought experiment. Close your eyes and imagine you are the energy being generated in the power station; take the journey through to the light bulb emitting light. You will identify the places you want to monitor.

The Sixth Sense

This Step, eyes, is really about observation. Observing data in such a way you can detect something went wrong.

Like we mentioned before, your eyes are just one of your senses with the ability to make observations. But you should consider using all five of your senses.

Smells, sounds, touch and taste, in theory, could all play a role in observing a problem.

Some people may have a sixth sense, a gut feeling they get telling them if something is wrong, but they don't know why. Neil deGrasse Tyson scoffs at the meagre achievement of only one other sense. He explains that in science, he has access to one hundred senses, maybe more. We have access to sensors and data that extend well beyond what evolution has provided to us humans.

In Engineering, we use these every day. We are also commonly the ones inventing them, in the first place. You can't (and shouldn't) always rely on visually seeing sparks fly to determine if there is electricity, you can use a multimeter to measure the voltage, amperage, resistance or continuity.

Every monitoring or measurement tool you are using to problem-solve are your extra senses - your superpowers to aid you in detecting a problem or pattern.

Your eyes in this step are both a literal and figurative tool.
We have so many tools available, make sure you are using the right ones and if it doesn't exist – make it.

Give yourself a sixth sense.

Your Checklist for this step

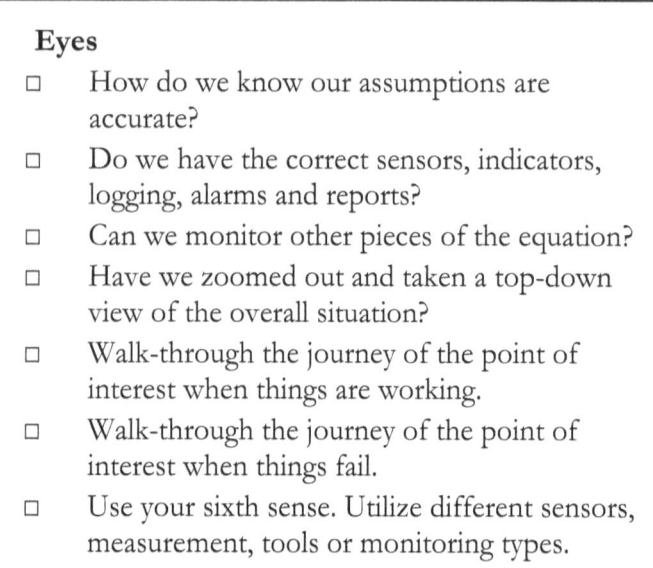

Eyes

☐ How do we know our assumptions are accurate?

☐ Do we have the correct sensors, indicators, logging, alarms and reports?

☐ Can we monitor other pieces of the equation?

☐ Have we zoomed out and taken a top-down view of the overall situation?

☐ Walk-through the journey of the point of interest when things are working.

☐ Walk-through the journey of the point of interest when things fail.

☐ Use your sixth sense. Utilize different sensors, measurement, tools or monitoring types.

If the problem remains, can you return to a previous step with this new data, or move on to the next step? Is there a chance that you may be the problem? – You better check your bases with Step 4. Check Yourself.

STEP 4.
CHECK YOURSELF

Before you wreck yourself

Have you tried turning it off and on again? Ugh, I hate hearing this question when I am dealing with I.T. support, because of course; I have done that step already.

So why am I suggesting it now?

Odds are, you are going to have to start digging deeper to figure out what is going wrong.

But one of the worst feelings is the moment of realization after spending a significant amount of time on a problem, and the fix was something simple you missed.

It can be quite embarrassing, but there's no need to feel this way – it happens to everyone. This step will help you avoid wasting time.

You might think, "Well, shouldn't this be covered in Step 2, the obvious?"

Yes, and no. The difference is, known problem fixes don't generally come with the same amount of time or effort as the types of things in Step 4, such as completely powering everything down and back up again, or rebuilding something from scratch.

This step is more than turning something off and on again. This step is about fundamentals. For those familiar with the OSI Model, fundamentals are addressing Layer 1.

What is Layer 1?

Consider for every bit of communication between devices there is a physical exchange - An electric pulse, a radio wave or some light.

A physical exchange is necessary for all data transferred over the internet. It still has a physical path to take – we call this Layer 1 in the networking world.

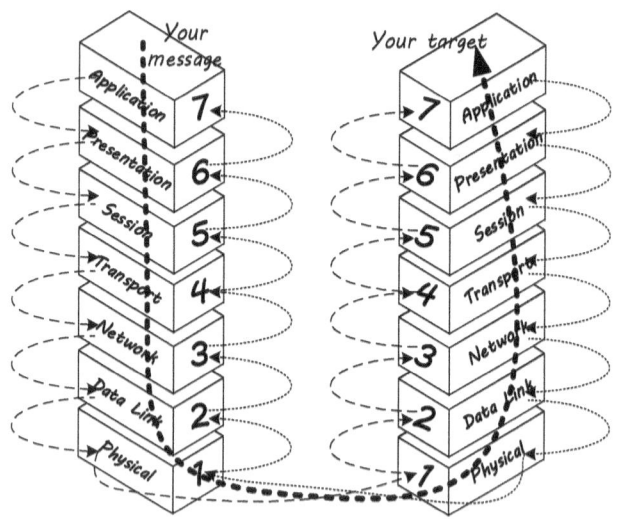

FIGURE 10 – MESSAGING OVER THE LAYERS OF THE OSI MODEL

The Open Systems Interconnection (OSI) model is a conceptual model to represent communications with a top to bottom functional partitioning.

Every time you send a message, update a social media status or make a streaming call, you are entering the information on the top layer called the application layer (layer 7 in the diagram).

The person receiving the message gets it by using the same application.

That message doesn't get sent that way. It is encoded down a few layers until it reaches the physical cables, sent over the internet, back into your modem and then up to your application and then ready for you to read.

Communication can take place at each layer. A user will typically only deal with the application layer, but without the lower layers, the top layers can't function.

Can you hear me?

This concept works in real life, too. Take the simple act of talking to someone. There's the message you say out loud, and there's the person who receives it.

If they didn't get your message you don't instantly start yelling, do you? No.

That's because it only solves the problem in one layer.

If they didn't physically hear you, sure, you might need to speak louder, but if you talk louder and they still can't hear? It could be a matter of them not listening to you, or they are distracted and not focusing at the moment. So this is the brain process layer, meaning speaking louder will not help.

Let's say you spoke as loudly as possible. Before a brain process can take place, it must receive a coherent message through the air.

A noisy area could distort my voice and interfere. A noisy environment would typically prompt you to move to a quieter place to speak.

For your voice to go through a medium in the first place, you must vocalize it in your voice box.

And before that, you need to take in some air.

Not the most elegant example, but the point is this: If someone didn't get your message after you spoke, you could put the effort in re-explaining it to them, phrasing it differently, whispering, raising your voice or adding hand gestures to get their attention.

But none of that matters if your voice box isn't working.

You thought you were speaking but were unable to create the correct vibrations in the air, to form the frequencies required to represent speech.

Examples of Layer 1 Checks

Some questions to consider if you're in the electrical or computer space:

"Is it plugged in? Is there a blinking LED? Is the power socket working? Is the power source at the correct voltage?"

Let's take it a conceptual step higher for other engineer types such as civil engineering, chemical engineering, biomedical engineering etc.:

"Are we looking at the right device? Are we looking in the right room? Is this the right variable? Are we using the correct tool? Is this the right building or street we are looking at?"

These examples might look similar to the last chapter because you are checking your inputs and your assumptions, but this step acts on a more fundamental level.

Even if you have logging and sensor readings all correct, you might physically be looking at the wrong thing.

In software engineering, this often happens as you are working on different versions of the same code. Almost everything is the same, but one thing is out of date which you thought you fixed. You do a bunch of re-work then realize you started editing the previous version.

Take a brief moment now to think about your field or industry. Ask yourself this question, "What are some fundamental, base-level things required for my work to exist".

Through your journey, continue to build up your list of simple stuff—a list of fundamentals.

Solving Application Startup Issues with 10-minute test cycles – Part 1

During project delivery, we found there was an issue with an application. After exiting the application, it wouldn't open anymore. The application ran on a windows server platform.

The best part was, we were doing a Factory Acceptance Testing (FAT) demonstration for some clients, and this was one of the last pieces of the demonstration. Of course, while keeping calm in front of the clients, in the back of my mind, I could only think

"What the eff is happening?"

I did Step 1 and changed the question.
Was it the way I double-clicked the application? What if I press the Return Key instead? Can I use the shortcut or the source execution file?

I then did the obvious in trying to open the application again. Exiting, killing the process in task manager, turning off all other applications, stopping windows firewall, antivirus software – anything with the potential to interfere with the application.

I progressed to step 3, checking the windows system logs, application logs, searching for clues.
The clients were still content, as had they experienced FAT demonstrations in the past. They were familiar in knowing they don't always go smoothly.
Everything was going well so far, so they weren't disappointed. The application worked well; that is when it was working.

I finally found a related error log, opened up the details and it read,
"There was an error starting the application" –

That's it.

I swallowed my expletive and just chuckled instead. The client asked what I thought the problem might be. I knew I was against the clock and I couldn't have the demonstration end on this note.

So I explained, "I'm not 100% sure yet, the best log I have is – there is an error – but I'll have to do some further digging to get a complete answer".

I needed to do something, anything, so I moved onto Step 4.

"I'll just restart the server real quick, just to get us going again to complete the demo and make a note to look further into this".

They were happy to be taking a step to move along, so I restarted the server. The fans stopped, you hear the system beep, the fans come on like jet engines, and we await the boot sequence.

At this point, I'm just engaging with the clients about general things, a little about what they liked so far in the demonstration, etc. What I didn't realize was the whole sequence was going to take 10 minutes to get the server running again.

The system booted up and needless to say, I open the application and *viola*, it opens with no issues, and we finish the demonstration.

While you might think "you didn't fix the actual problem", that is partly true, but you have to remember we had two problems to solve.

Problem 1. Why does the application fail to open after it is closed?

Problem 2. How do we get the application started to complete the demonstration?

Before creating an analysis of all the possible root-causes, devising solution plans, evaluating and selecting something to try, I did Step 4 check yourself.

Restarted the operating system and tried again.

I completed the demonstration with both the client happy and our management happy, promptly in a higher pressure situation. So problem two was resolved, with just problem one remaining.

I did end up getting to the root cause of problem 1 – in **Step 7, Strip**. I'll make sure to continue the story there and show you how it ultimately came to be solved.

The section title is the same as this, **part 2**.

Turn it Off and On Again

There are some cases similar to the story above, where the problem never resurfaces again. So, digging deeper into why and how it happened in the first place ends up not being worth it since it would require such a significant amount of time for little payoff.

If we are already using the application and no one is reporting any further issues with it, then why fix what's not broken?

It is a red herring, an anomaly, a spurious event.

Anomalies are the reason this step occurs at this moment. We don't want to start with it because it is time costly and using secondary reasoning, but after completing Steps 1-3, as you can see, it is perfect timing to consider this step before moving on.

The type of problems that get solved with this step often feels a little embarrassing if you spent any longer doing the later stages, the more time you spend, the increase in embarrassment. It's a linear relationship.

Don't get me wrong, I'm not talking about the judgement of your peers, although this may happen. It's to address your judgement of yourself, for not covering your bases. So do it now.

Check yourself before you wreck yourself (emotionally).

The Difference between Something Obvious and Fundamental

At the beginning of this Step, I did touch on the time differences between the obvious and the fundamental.

In the application story, a Step 2 "obvious check" was to kill a process and to stop and start the application.

These two tasks may sound similar to the action of restarting the server. Why can't turning it off and on again be considered an obvious step?

The answer is, it can be.

To know when it is one or the other comes with experience and specific situations, but one other distinction is to think about it in the OSI layer model.

A restart at the application layer was the top layer and is why it is typically akin to the shortest time. But another reason is that it has the least dependencies.

Other applications sit on top of the operating system, so when we restart the physical server, all applications get restarted.

Obvious step 2 type known fixes, typically sit at the top layers and don't have other dependencies while fundamental ones impact the lower layers.

The Problem the Whole Time

I knew a student who was working on getting a full-bridge converter functional. It's a typical configuration with four active switching components across a power transformer.

With a full-bridge structure, the output voltage is equal to the power supply voltage.

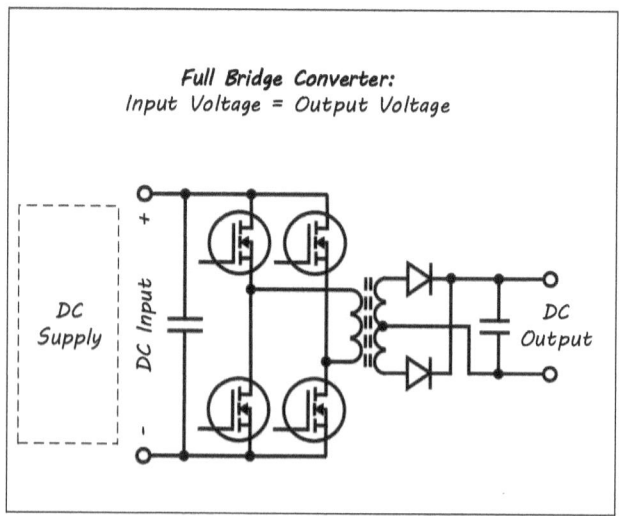

FIGURE 11 – FULL-BRIDGE CONVERTER

So he would know he set it up correctly by measuring the output voltage (DC Output) and confirming it matches the power supply voltage, 12V in this case.

The problem was that the output was displaying 12V but as soon as he ran the circuit, it would fail.

Some advanced testing could include checking waveforms or simulating different inputs and responses. They decided to go down the rabbit-hole of testing different versions and combinations of those techniques.

A few hours went by and still no results. The thing is, this wasn't the purpose of setting up the circuit. There were a few steps to set up the lab test, and the first one was setting up the full-bridge converter.

After setting it up, he needed to use it to interact with other circuits. Imagine spending the majority of your overall task time on the first step and making 0% progress.

It would start to get frustrating.

Using the principles of this book, the tasks undertaken to investigate further would happen in Step 7. In step 4, Check Yourself, there's an opportunity to look at fundamentals, or layer 1.

So what is layer 1 in this context?

In the electrical engineering world, the input supply is a base dependency for everything else in your circuit to work. The same way no electronics work unless you have a working battery.

Like when you're about to start gaming and the controllers have no charge!

In this circumstance, the issue wasn't as obvious as a missing battery, but the problem did end up relating to the power supply. It turns out that it had built-in current protection.

The current protection feature meant that while the circuit was running, the protection would turn the circuit on and off.

It turns out that he set up everything correctly initially, but he undid his correct work while troubleshooting, and began doubting the solution.

He used a different power supply, and all of a sudden, everything worked.

Risks When Changing Fundamentals

It's not always a no-brainer to mess with the fundamentals of a situation. The risk you have to consider is not to introduce a new problem.

Let's say you do have to restart a server. You may impact services on another machine, client or server that you now have to solve too. They become your problem. The probability of this isn't too high as you are already in a broken state and trying to fix something, but it is worth considering.

We must avoid adding unnecessary variables to the situation. The way to achieve this is only to change one thing at a time. If that's restarting one device instead of three at once, with some configuration tweaks, so be it.

The reason is, if you do end up causing a change in behaviour, which is another clue – you cannot attribute it to a single thing.

You don't know which of the three things you changed at once actually helped.

Sometimes it is good enough to make it work and is why I say avoid doing it, rather than advising never to do it.

The traceability is what you need to consider.

CHECK YOURSELF

Your Checklist for this step

Check Yourself

☐ Identify what Layer 1 is of your problem

☐ Can you restart a system on layer 1?

☐ Check if the device is connected.

☐ Are looking at the right object or device in the correct location.

☐ Have you turned it off and on again?

☐ Did we consider replacing the medium? (E.g. cables, power supplies, etc.)

☐ Avoid adding unnecessary variables to your situation.

Checking these fundamentals will save you any headaches or embarrassing explanations. It will reduce the odds of spending too much time on the problem.

At worst, if the problem remains, we can now return to a previous step with this new data. You can even try repeating the obvious fixes or seeing if the eyes you set in place have new information.

If pain persists, see your doctor, in the next step, Doctor G.

STEP 5.
DOCTOR G

Let Me Google That for You

There's the time and place for asking a question. In Step 1, we established the importance of asking better questions and asking more questions, while here at Step 5, I am telling you there is a time and place for it.

Specifically, you should be clear in what you are asking and then understand if you have a question for Google and not for a colleague or another engineer.

In this chapter, we try to understand which questions are right to ask others and which questions are better to direct to a freely accessible public repository of information. We also learn how we can do this effectively and quickly, finding more data in less time.

If you have had problems, where you only received vague answers or caused some frustration, this step will help you know when you should be asking these questions in the first place.

How do I know when the right time is?
The clearest indicator is if the question sits in the public domain, or if it is at a commercial product level.

If it requires a subject matter expert or is a specific proprietary product, you will have limited results.

Alternatively, if the method or technique used in whatever you are doing is under some patent or copyright protection. The odds of you find something is also limited.

But this step isn't merely "googling the answer". It's more about reaching out to the publically available information on the topic. This outreach could mean joining a forum or group who are implementers, or other engineers in the field discussing questions less formally.

Which groups can you join about your field of expertise?

GOOGLE IT.

Okay, I set that one up, but in all seriousness, there are questions worth asking and others where it takes a fraction of the time to search. If something is a pure memory task and requires zero analysis or expertise, then it's a job for computers. Otherwise, it is appropriate to ask.

Dealing with Vague Problem Statements
The reason a senior engineer's answer might be vague is; first, the lack of time, they gave you the task not

to walk you through it and spend the same amount of time training you, but for you to make a problem go away. Their ideal situation is they spend zero additional brain cells on the issue.

The other is they may think something is clear and so they give you a too high-level answer that is probably true, but you don't even know what those string of words mean in this context, like the story of my first task in the framework chapter. The job is clear as mud.

Let's say your manager gives you the task of adjusting a temperature warning set-point for a custom electrical panel.
Your job here is to decide at what temperature the panel will send a signal to inform an operator it's getting too hot.

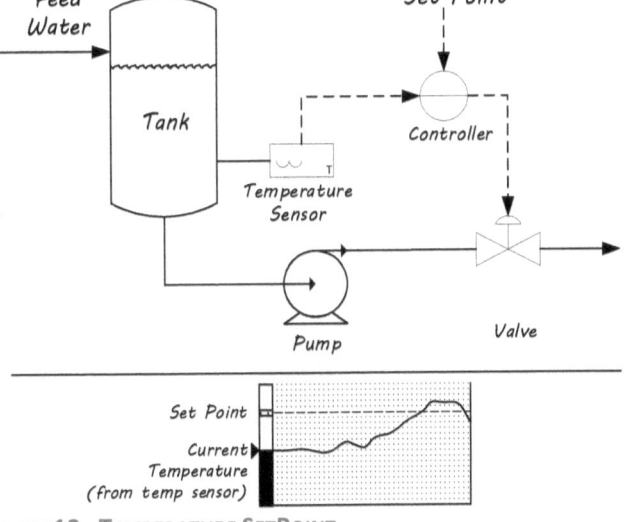

FIGURE 12 - TEMPERATURE SETPOINT

For simplicity, we can break this task down into three main questions:

1. How do I adjust the set-point?
2. What is the recommended operating temperature/ standards of the electrical panel?
3. Are there other electrical panels I can use as an example?

The Single Worst Question you could ask

Now, it is tempting to ask the question "what value should I set the setpoint to, and how do you want me to do it?"

This formulated question may be the single worst question you can ask at this point. First, your manager likely gave you the task because someone else couldn't do it themselves in the same time frame.

Secondly, suppose you consider another engineer as a finite resource (due to time) over the lifetime of the problem you are solving. In that case, you may only get a few interactions to have your questions answered.

You want to make sure the question you ask is worth asking. Now I know there are no stupid questions, at Step 1, you would have asked these types of questions already.

Maybe you've never been trained on this topic or exposed to it. But bare with me, at this step, you want to ask "the next question".

The question you cannot get answered by Dr Google.

Let's do an exercise to see which of these questions you should ask google and which you should ask a more senior engineer.

We will use the temperature setpoint scenario for this exercise, and you can refer to the diagram if need be.

This is how it works, of the three main questions listed above, can you rank in order which ones are the most google-able to the least?

Ready?

Here are the questions again:

1. How do I adjust the set-point?
2. What is the recommended operating temperature/ standards of the electrical panel?
3. Are there other electrical panels I can use as an example?

The only answer is 2, 1, then 3.

Question 2 is the most google-able because the recommended operating temperatures and engineering standards are not secret information and are designed explicitly for reference by engineers.

How do we know which is google-able? Remember, if it is purely a memory task and requires no analysis, then it is likely a question for Dr Google.

Next, for question 1 "how do I adjust the set-point?"

I didn't specify if this is a proprietary product or not, in either case, you can google this. If the product is proprietary maybe check a specialized vendor forum or equivalent.

You might just find an example or instructions.

Okay great, we have potentially got the answer to "what is the recommended temperature" and "how we can set it".

The last question – do we have one already in the field as an example? You won't be able to find this on the internet, so you may automatically ask this question of someone else, and it wouldn't be a bad one. But there might be another resource to refer to before asking.

Here's a clue – it's not an online resource.

What about literally going and checking yourself? Assuming it's safe and accessible, you don't need to ask at all and can check for yourself – more on this in Step 6.

If this isn't possible you should ask this question, sure.

What did we learn from this exercise?

We turned four back and forth questions into zero and at worst two. You solved a problem in less time and fewer interactions than the average engineer!

Before you complain about the lack of training, the vague direction, the flawed personalities and the high expectations, consider this approach.

You might be right, some companies and personnel have more issues than others, but guess which of these are the most in your control?

Only your actions.

Which means you can more consistently repeat this step compared with waiting for your company to organize a training session or your senior engineers to have time to sit and explain things, or for them to grow a better attitude (this one's a long shot over a lifetime).

Using this technique, your new first question to the senior engineer might be:

"I've selected this set-point based on the engineering standards.

I also checked another panel and saw we set it to this value before.

I will set it using these instructions I found here.

Can you review and let me know if okay to proceed?"

Comparing with the original first question:

"What value should I set the setpoint to, and how do you want me to do it?"

To the untrained eye, the original first question is a seventeen-word question, compared to a whole paragraph using the new technique. A shorter question should be the better question to ask, right?

Wrong.

What did we say at the beginning of this chapter?

The task is given to you so someone else can spend their brain cells elsewhere.

By presenting them the simple question, you are asking them to answer five others first. If they walk you through the task, they may as well have just done it themselves.

You didn't achieve the goal of saving them time, nor did you save them brain cells, and the problem is still there.

With the longer paragraph, although more to read, it has three key points they can look over, and if correct, they will say "go ahead". If wrong, they can address the specific incorrect part and move on.

If it's your first time, it is understandable, of course, but just before you raise your hand, please breakdown which questions are for Dr Google and get that out of the way.

Get your ears close to the floor

For you to be effective with Step 5, it means you need your ears close to the floor.

What do I mean by this?

You need to find the groups and forums related to your field of expertise, the podcasts, the films, and the books and absorb them, be around the conversation, so you have a resource to consult. Google isn't the only online repository of information on the internet. Be prepared, not lucky.

"But if I spent all my time googling instead of asking someone experienced, doesn't this take a long time to research versus being explained to?"

Sure, you could ask straight away and get a quick answer, but you have to be aware of why they gave you the task.

One other thing to consider is the duration of the quick answer may turn into a 5-minute conversation.

Let's say, without asking someone, you have to analyze the information and figure out how it applies, and it takes you 30 minutes, compared with the 5-minute conversation.

This pace leads you to believe getting the answer in five minutes beats 30 minutes of research—a fair assessment.

But this doesn't mean you don't do this step. It means you need to get better at this step. Don't forget that you only counted your five minutes. That math isn't quite right. You now need to include the five minutes of another resource. So five minutes of your time ends up being ten minutes of project time.

We haven't completely closed the gap so, what if you could more effectively research a topic rather than 30 minutes, you cover the same issues in 10 or 15 minutes instead?

You can reduce the amount of time researching a topic by making sure you google the matter effectively.

How do you google a topic effectively?

The biggest problem I see for most people is they dawdle when they look things up. Here are some tips for you to make sure you are using search optimally.

The 4 Elements to Faster Knowledge Downloading
When describing the brain-computer interface as a type of AI symbiote cyborg, Elon Musk often points out

that we are already cyborgs when you think about our usage of the mobile phone.

It's just that we have minimal bandwidth between our brains and the information on the internet.

When thinking about downloading knowledge from online, there's surface-level knowledge, and then there's a deep understanding of the topic.

In most cases, all we need is the surface level knowledge and then leverage the below keys to go deeper. Here are 4 Key Elements to faster knowledge downloading from Google:

1. Power sift,
2. Results as questions,
3. Media spam,
4. Focus switch

You will consume more data and better focus your queries in less time by doing this.

I'll walk you through a typical scenario where someone gives you vague instructions and then requires a search task to resolve the issue.

After, I'll expand this description and show you how you can use the four key elements.

A Normal Search after Vague Instructions

Imagine you work for an electrical company. Your company has technical diagrams it generates automatically from data in a spreadsheet.

There's an adjustment needed to change where data is stored.

There is a button in the spreadsheet they press to trigger the process. In this scenario, you have some basic coding skills and can work your way around a computer – but you have limited experience with the capabilities of these programs.

Another engineer gives you a vague task, and they ask "Can you please adjust the program that generates the Visio diagrams from Excel?"

"Which program?" you reply.

"The main one we use, don't worry, you can access it from your laptop, just use the shortcut on the desktop" she explains and then leaves you to it.

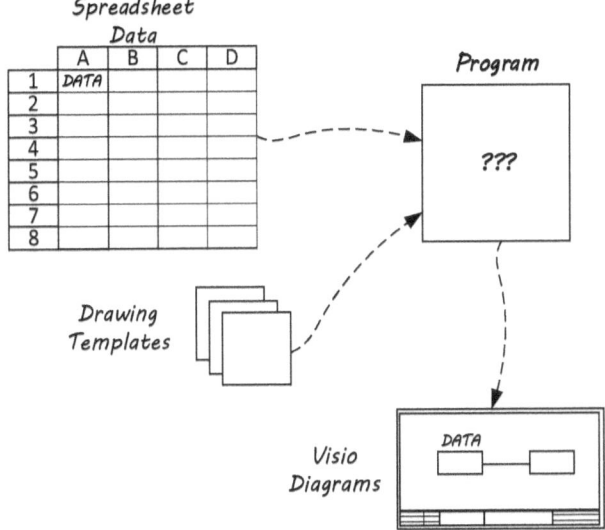

FIGURE 13 - VISIO GENERATED ENGINEERING DIAGRAMS

So naturally, you google "how to program a Visio diagram from excel".

You then click the first link, read through, go to the next, and read through again. Fifteen minutes later, you've found a bunch of unrelated information. (I checked, there's some cool stuff but it isn't this scenario).

You spend another five minutes re-typing your question and another 10 minutes reading those articles.

In 30 minutes we have nothing. We have checked about six articles and tried three different queries.

We know a lot about data visualizers and all sorts of stuff, but the task is too vague! We wasted a bunch of time and need to ask for help.

The 4 Elements

Earlier I mentioned the four key elements. Using this scenario, I will talk you through using them to save you time and more likely arrive at your answer.

The first 15 minutes of reading through articles should be about a 1.5 minute task – 10x faster than the first time.

How? By power sifting.

Power Sifting Query Results

Let's assume the same query into Google. Instead of clicking and reading through, I want you to read the title and open in a new tab in the background, (without navigating to it) if it seems slightly relevant for about five tabs total.

Now go through each tab and using headline reading only (the big, bold headers) determine if it seems relevant to what you are answering, if not. Close. Next tab. Spend about 15 seconds per tab.

If we assume 15 seconds for you to go through the list, it adds up to a total of 90 seconds.

If you feel like it is close to the answer, feel free to repeat this step going further to page 2 of Google, and within a few minutes, you might find the article you need.

If not, we rely on the next element – results as questions.

Using Query Results as New Questions

Consider that the people writing answers to questions didn't write these answers from the same angle you are trying to write your question. The language used in the answers you found so far will lead to better queries to make.

We have now closed a majority of the tabs because of limited relevance, but they had some keywords as well. Use these keywords to form your next query.

On the 3rd page or so of this query you start to see some results about Macro's, you then "power sift" and find another article talking about VBA. Next, you change the original question and add the keyword VBA, and now you see a result like "create Visio document from excel using VBA".

The results say you can create buttons, and now your query is a lot better. You find specific articles.

So in 3 cycles, we found a useful article. If you're keeping count, we're up to 5 minutes.

What if we didn't find this convenient correlation? Then we expand our search by performing media spam.

Utilizing other forms of Media Results

Search engines are returning articles, but you can also open up a few videos, images, shopping – you might find a product-related ad.

They have the phrasing to sell the product which you now use as new questions and repeat your query using these better keywords.

You see the pattern here. The videos people make are there to try and answer questions, so content creators try to find what people are asking for and create content around that, you can take advantage of this by seeing what they are trying to focus on in their keywords.

Don't forget that Youtube is the second largest search engine, right behind Google. You're an engineer – you know this to be true.

The video might just explain what you need or have a URL for what you need. Sometimes expanding the video descriptions have useful keywords or links and better yet, sometimes people in the comments section ask related questions (you can also get a laugh from the ridiculousness).

Set videos to 2x playback speed and click through to see if they are answering your question. You can digest 10 minutes worth of video content in about five minutes.

If this fails? We are at about 10 minutes now.
The last element is the focus switch.

Switching Focus for Finding Answers

You've spent so much time coming from one angle, let's look at our task and start our queries from the other end.

"My excel button stopped generating Visio diagrams."

Not the best set of words, but instead of searching how to do something, you assume it's there and instead, you ask how it gets fixed.

We repeat the above techniques for the new query, and this takes us to 20 minutes.

So what did we achieve in 30% less time?

- **Basic Searching:** 6 articles and three queries
- **Using the 4 Elements:** 30 articles, six queries, 20 minutes of video/audio content.

That is a 500% increase in text data, a 200% increase in queries and infinite increase in video/audio content, all in 30% less time.

You might be a little sceptical at this point because the math I just showed you isn't super scientific. These are arbitrary numbers, and the scenario is anecdotal at best. But the purpose is to paint a picture around the type of proficiency you should possess with universal tools such as search.

If you type slow, learn to touch type, if you struggle to read your screen, increase the DPI settings, if you can't sit comfortably for long, adjust your entire work set up to maximize your capabilities.

STEP 5.
DOCTOR G

Imagine you are a professional gamer about to compete in a tournament, you would make sure you have the best equipment to reduce any lag and make you comfortable. You are a professional engineer who is ready for problem-solving action.

The reason I am pointing this out is that they are all bottlenecks of the data transfer from the source information to your brain. Reduce these bottlenecks, and you will be surprised at how much more competent you can become.

Your Checklist for this step

> **Doctor G**
> ☐ Is there a part of the question available online?
> ☐ Have I used the 4-elements in searching?
> ☐ Are there other forums or online communities that specialize in this field?
> ☐ Have I gathered the data I can by checking myself?

Becoming a better "googler" or "query asker" is a skill that will let you churn out answers faster and ask better questions in the first place.

Think about all the times someone asks a vague question, and it is because they don't know what they are asking, exactly.

In your interaction with them, if you came up with a few answers for them depending on if they were asking one question or the other, they will be able to refine their query for you (or pick a path you laid out).

Some engineers enjoy working in this way as if you are the auto-complete to their sentence.

If the problem remains, we are starting to get into deeper waters here, but it also means if we've got nothing, maybe we are asking the wrong question and can return to step 1 and make our way back here. If you're sure you have the right question, you might need something a little more substantial. Sometimes the answer is not on the internet, then where can you check next?

Move on to the next step to learn how to deploy the RTFM Protocol.

STEP 6.
THE RTFM PROTOCOL

Read the Effing Manual

Have you read the manual? Don't scoff at this point. It's a real consideration!

Engineers are good at figuring things out and are notorious for not checking the manual. If you are up to here, it means you have exhausted online means of searching, and there might be one specific source of information left, which is the manual.

Other documents to consider are datasheets, accompanying information with the product, reports, specifications, statistics, etc.

If your situation or problem has no associated manual, this doesn't mean this step is over. What you will learn from this chapter is using the information to serve you.

You learn how and why taking this time to stop and absorb tangible information is powerful.

Do your homework.

This Step is where you start worrying about planning and preparing ahead of time. It's the first point where we stop and preplan options.

This step exists because you can find common troubleshooting steps in these manuals specific to the system. Also, I repeat, engineers are notorious for ignoring the manuals.

Think about it this way: when another engineer might have developed this product, they had a list of defects to handle. With some of the problems, rather than fixing it directly, they "fix" it by documenting the troubleshooting step as a workaround. It takes them less time than finding the root cause and gets the product out to market.

This situation could go for particular materials; it may not meet exact character profiles. Still, when using the material they've selected, they list the conditions where you can expect certain behaviours as advertised. So you get the datasheet.

As a secondary help, you may not get the specific answer, but they can provide more clues as to how detailed the manufacturer of the product or tool was, or how mature they are in their development process.

If you are keeping track, you may notice that at six steps into the 10+1 process, you might be starting to creep towards the "expected time to solve" line.

Time is not on your side, and you need to prepare to try more challenging or risky things—the ones you've been holding off trying before this point because the effort is relatively high. There could even be a risk of downtime or impacting others from doing their work.

We use the RTFM step not just to read the manual but also to write the manual, per se.

Begin to articulate what you have done so far, what hasn't worked. What are the new suspicions and what we want to try next – continue the narrative from step 1.

Write the next five steps of what you want to try so you can begin to document the method and any variants. People accidentally do this step earlier in the process, like in Step 2, the obvious.

The idea is to develop an investigation plan. One that you can set up to test different parts of your problem. Each step intends to gather data.

The beauty of this approach is that even if you end up at this point, the things you've tried are no longer assumptions, they are your starting data.

Combining this starting data along with the data from the results of the investigation plan in this step, should get you closer to finding the problem. It will serve you through the rest of your problem-solving process.

Pro-tip; if you have two people, you can have one experimenting towards this step while the other tries to explore steps 3-5 in parallel.

The reason why you only do the obvious checks and not iterate endlessly is that these further tests inherently take more time and come with higher risk.

It is appropriate to do these riskier checks if you have reached this far in the problem-solving process.

How do we mitigate the risk of rigorous testing?

Before I introduce you to the 8-step investigation plan, we need to understand the risks. The risk is about the things that could go wrong on the existing system, product or situation when performing troubleshooting tasks.

You must have all the failure modes and what your response will be before performing each task.

It is essential to know that you are coming from a "brute force" mindset where you are trying a set of things in a small timing window.

Similar to how a hacker might brute force guess a password where they try every combination of characters.

If a) works, okay, if a) fails, try b). If b) fails, try c).

You record the results and make sure you know what you want to record. Taking note is vital for reliably creating conclusions and potential options for the next step.

It has the added benefit of using this to report to your stakeholders or upper management so you can avoid the high-stress zone for a little longer.

Don't try to fix the problem now; just gather the data and move on.

Handling a High-Pressure Situation Needed for Testing

"Are we shutting down?"

The client is standing by ready to disconnect the system as soon as I give word things are going wrong.

I am monitoring the situation via a computer screen, which reads an error:

```
Request timed out.
Request timed out.
Request timed out.
```

There is a low whirring sound of giant machinery, which is comforting because it means everything in the power station is still running.

I'm working on a live control system network. If I make a wrong choice here, the best-case scenario is to re-assess everything and start again, explaining to several people what went wrong, and the worst case?

Blackouts. Thousands of homes lose power.
You know, nothing too dramatic.

Okay, in the planning, we knew this was one of the potential outcomes; I have at least 8 seconds before making the call to shut it down.

How does one solve a problem under such circumstances?

You guessed it - The RTFM Protocol.

Before this moment, I had already done my homework. Let me clarify. The situation required a live cutover between an old system and a new system.

Live means the system I am working on is running and controlling some big machinery.

Imagine you're driving your car in the middle of a highway at 100km/hr, and you just need to switch the engine over to a new one, but you can't stop the car from driving. If your engine stops and your vehicle comes to a screeching halt, you could get in an accident. So the equivalent is you swap the engines while driving at full speed instead.

An 8-Step Investigation Plan

To keep things simple and about the process, I'll just briefly explain the plan and point out the tips you should consider when managing a situation like this.

The approach for testing goes something like this investigation plan:

1. Identify the tests you want to do.
2. List what behaviours you expect to happen.
3. List the behaviours you don't expect to happen.
4. What are you going to record to prove the behaviour? (eyes)
5. What is your next action IF the behaviour is what you expect?
6. What is your next action if the behaviour is not what you expect?
7. What external things are at risk by your actions and how are we monitoring them. (eyes)
8. What is the fallback plan if an external thing reports a fault?

THE RTFM PROTOCOL

With these answers in place, you can have some confidence in your test. The part where it gets intense is when you have follow-up tests for different results.

If a) works, okay, if a) fails, try b). If b) fails, try c).

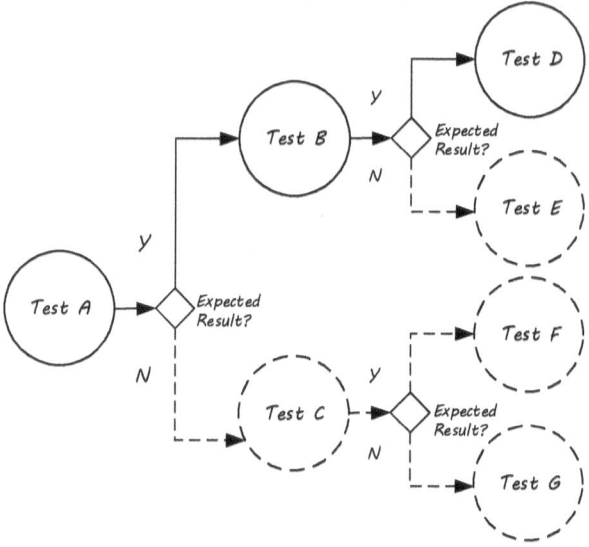

FIGURE 14 - TEST PATHS FOR INVESTIGATION

Prepare another list for approaching the test, and you will find you can rapidly satisfy those answers because they will be similar to the first. You use the same set of eyes to make sure you see everything to determine if your predictions are as you expect.

You try having a similar fallback plan in place. The most tests I have done to date is about 47 of these in a row. I was deep into a problem.

So, back to the situation in the control system network. We have a fallback step for shutting down the test by removing a device's network connection if the behaviour was unexpected.

The client has his hand hovering above the cable port. If the network remained in this state for a certain period, one of the other systems would produce an error – a 10-second time out I pretend to be 8 seconds to give me a couple of seconds of contingency.

I was about to give the call to pull the plug, but I had eyes on another system which wasn't reporting an error, so I had the confidence to stay the course. I collected this data, and I already had the next two tests to try lined up.

After three tests, I had the information I needed to know what we should do by returning to step 1 and then coming back here to step 6.

Yes, you can jump between steps like this.

To see an example of the testing sheet I used in my engineering notebook, you can refer to the "Rapid Test Sheet" instance in Step 7.

What's the value of documenting the chain of tests?

You can come to a solution faster, give your clients or your senior engineer's confidence because you have a plan.

You automatically manage their expectation by explaining these are investigative tests and not the trial of a solution.

This way, if all you achieved was acquiring data without causing interruptions, they would be happy with the progress.

It would be considered a win.

Compare this with if you set the expectation you are trying something to fix it and then at the end of your plan, you don't have the solution, they would look at that negatively.

This type of plan and approach gives:
- A faster path to finding solutions
- Lots of data to work with
- A way to show initiative and preparedness
- Confidence to stakeholders
- A means to eliminate possibilities
- A compass to navigate between several test results

Remember this; we either solve the problem or gather more data.

Control Logic – False Positives

During a month-long commissioning project, we were connecting Input/Output (I/O) devices from the field into our primary control system.

To commission the site, we connect the wires between the field device (a fan, light, valve or motor) to our Programmable Logic Controller (PLC). We are essentially enabling our controller to turn the device on and off.

We then test each control in the device, and if it behaves as expected, it gets a tick, if not we go and find out where the error is. Typically the device has an issue, or the wire is damaged.

We are looking for the press a button on the screen which correlates to the action. If I double-click the on switch for the light whilst in the control room, the field engineer should be able to confirm the light is on in the final location (It can be more than 100m away).

When it fails, we troubleshoot.

So we had a particular I/O point which was giving us a headache. We went off and performed all the obvious checks, trying it again (Step 2). We got some eyes on the situation with the field engineer using a multimeter to measure the voltage at the device end (Step 3). We also disconnected the device, so we were only checking the control signal, covering our fundamentals (Step 4). The signal was okay, and there was no visible damage.

This particular issue had dragged on longer than it should have and finally, someone asked "Did you check the contact definition" to which our tech replied, "yes, the logic has two outputs, light switch on and light switch off."

He was correct the logic did display this and the control system was sending an on command on the right output. The second output was false.

"But did you actually check the datasheet of the field device?"

It turned out that the device was a Normally-Closed switch and not a Normally-Open switch. Typically, you expect this type of device to be normally-open, which means when you send a positive signal from the Controller, it closes the switch, therefore turning it on.

But as this was normally-closed, every time we were attempting to turn the device on, we were turning it off. Effectively making our following of the labels have the opposite behaviour.

The lesson?

Don't rely on labels others have placed, read the datasheet (Step 6) - RTFM.

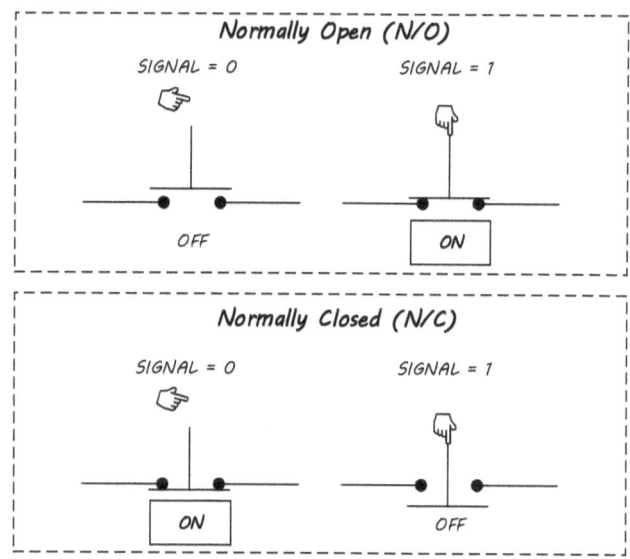

FIGURE 15 - NORMALLY OPEN VS NORMALLY CLOSED

Open The Box

In a job interview for a technology retail store, I got asked "You're in the store and approach a customer.

If they asked about a product feature you had no idea about, without checking the internet, looking it up or asking someone, how would you find out?"

Naturally, I rambled through a bunch of options that were still variants of asking someone else. For clarification, you couldn't see inside the box, and the product details on the box did not have the information either.

They narrowed the question down "The box is for a toy, and they wanted to know if it came with a specific adapter".

In my mind, I thought if it's missing as a feature on the box, we can likely assume that the manufacturer didn't include it because they would want to show off all their features to potential customers.

So my answer was to explain to them I don't know off the top of my head, but based on the information we have, I wouldn't assume it had it since they didn't advertise it on the box. Then offer the customer to go and find out if they were still curious.

This answer was, okay.
The best answer was-
"Open the box".

I thought, wait, am I allowed to do that? But let's be honest. I didn't consider it because I was worried about the rules. Otherwise, I would have asked them if I'm allowed to open the box.

The point is, sometimes you have to go and check for yourself. Maybe it's not on the datasheet or the manual, but if it's something you can go and look at, check – similar to what we said in Step 5.

STEP 6.
THE RTFM PROTOCOL

So what have we learned in this chapter? We learned reading the manual is valuable, and you might find the answers there.

We know reading the manual and writing a manual for your investigation plan is recommended. By writing, this manual, we have all the failure modes considered, and we know where to want to plant the eyes for the investigation.

With this plan, you can build a chain of tests to undertake to attack the problem and give you as much data as possible with slightly varying circumstances to analyze later.

We also know that writing things down helps our brain process information and checking information as close to the source as possible can be helpful, particularly if the information was an old label or document written down by someone else in the past.

In some situations I've been in, someone from a company that doesn't exist wrote the source information.

The icing on the cake?

They wrote it over 20 years ago, and they don't even make the product anymore.

We all love good documentation. You even curse the person who left you with poor documentation. So use this step as an excuse to start documenting the solution better for future you.

Your Checklist for this step

The RTFM Protocol

☐   Have I read the manual?

☐   Have I checked all supplied documentation?

☐   What tests will give me more clues to the root cause?

☐   Have I made my 8-step investigation plan and answered those questions?

☐   Open the box and check

It is a typical case to return to one of the earlier steps and arrive at an answer after getting information from this step.

If not, it may feel like you are looking for a needle in a haystack and running out of options.

You may start to "feel the heat" but fear not, there is still 4+1 Steps up your sleeve you can try – with temperatures rising, the next step is to strip.

STEP 7.
STRIP

Reduce Complexity

Remove a layer of complexity or a variable from your scenario and prove things work at each simplified layer.

We are starting to get deeper into our problem-solving process and have checked and tried a combination of things in the previous step. This problem is a tough one so let's change the playing field.

This step is about shifting gears to give you some confidence back now you have been defeated a few times by stripping back the complexity.

How far back do we strip?

We strip as far back as proving 1 is equal to 1. Build some confidence that we understand our place in the universe. Okay, not as philosophical, but at the very least we know where we stand with all our inputs and assumptions.

In a software environment, you might have an IF statement:

IF <u>current temperature</u> > 5 degrees THEN raise the alarm.

At step 7, we might want to check something as silly as forcing the current temperature to be 6, so we know for sure the alarm is raised, because 6>5, right?

Next, we want to prove to ourselves, that yes, the temperature of 6 is greater than 5, and we are getting a true statement. *Prove 1 is equal to 1.*

What about in other forms of engineering? Instead of proving a software statement, you want to establish a principle is true, whether it be one of physics, chemistry or math.

For the problem you are working on, what are the core principles or universal truths? Which laws do you need to consider? What's a statement that is always true?

For example, in Chemistry if you were to combine one atom of oxygen and two atoms of hydrogen, you produce water, not sometimes, *always*.

Now if you don't get water, something is completely wrong.

Remove everything until you can say a true statement or a principle, and prove the situation works in this adjusted state. Pull back the levels of complexity until you have certainty on where the uncertainty begins.

The Physics Approach

Elon Musk always talks about the Physics Approach when it comes to designing the right solutions. This approach means starting with physics, with the base knowledge.

The electrical pulses representing data, the math behind a signal, the molecules of your material and reason your way upwards from there.

Now, we are not always designing a brand new product or solution; we are usually trying to fix a problem with something existing. You still use this same thinking.

The physics approach is essentially going to Layer 1 and reasoning up from there. If you recall in Step 4 Check Yourself, we are looking at the fundamentals or the layer 1 components of our situation. Here in Step 7, what you are doing is looking at the OSI model again, but this time, all seven layers.

If the OSI model is not directly applicable, then whatever your highest layer is will be the one with the least dependency, as defined in Step 4. You considered the top interaction where the failure occurs and all the dependency layers underneath.

To do this step, you are just removing one variable from the existing layer, or you are removing a dependency and checking at the next layer down.

This layered approach is systematically checking each layer one by one. You cannot skip straight to the bottom layer (we already did this in step 4) as you won't be getting the benefits of the exercise itself.

You should be able to make predictions at what the new behaviour will be as you remove layers of complexity. Remove layer 7, what will happen? If you then remove layer 6, what will happen?

Hacking Critical Infrastructure with a Printer

I'm on-site and just had the morning safety toolbox meeting.

We discussed all tasks each team member will be doing and the areas we are working in, confirming we have our appropriate personal protective equipment (PPE), skills and documentation to carry out the job.

We look at if we may be impacting others or if they will be working within the same work zone—a very productive meeting while drinking some very welcomed morning coffee.

I look at the first task in my list of thing. Then I get on my safety gear and head out to the location to do the work. It's a quiet room, so I'm not expecting any interaction but all of a sudden, the client approaches me with a concerned and urgent look.

"What are you working on right now?"

I had just got to the location and hadn't started any work, just reviewing the steps and doing safety tasks.

"Did you change anything?" he pressed.

I said, "No, not yet, I'm just going through the work I've got scheduled".

He was stern now. "I need you to come with me".

What I didn't realize was my response was considered evidence that I was potentially not telling a full story.

He was there in the first place because he got information about a change that occurred per my request the day before, but I could not see how this was of concern.

I had requested a network port to be assigned so I could connect a network printer for our team to use, which is a typical requirement for a project team to have—the ability to print documents.

Given the nature of our interaction and the fact I packed my belongings to discuss further with management, was confusing. This predicament was the time for being observant, getting all the information and reserving reaction until this point.

We get into a meeting joined by my site manager to discuss why I shouldn't be escorted off site right away.

I won't get into too many details except for the problem-solving portions. Still, the premise was since I am an OT Cyber Security Specialist, I have the capability to hack further into their networks. By requesting this port access and connecting a network device (the printer), I now have the means to do so as well.

There are two questions – one about governance and compliance; did I follow the proper procedures, inform the correct people and stick to the policies.

The other question is about technical feasibility. Let's stick with technical feasibility.

Could I have *really* hacked this corporation with a printer?

Okay, I phrased that in jest, but it is a complicated question nonetheless.

That's because this was a "Loaded Question", remembering what we learned about the types of questions in Step 1, meaning answering with a simple yes or no is ill-advised as I will be saying yes or no to more things than it seems.

Although there may be a minuscule possibility of a successful cyber-attack to occur, I intended to show it would be quite complicated – almost impossible, even if I wanted to do so.

Therefore, I treated it like a problem and stripped it back to the point where it could occur.

The following is how I explained it by creating a chain of events sequence.

Using the Strip Method to Create a Chain of Events

The fundamental, figurative "Layer 1" of this situation is; I gain unauthorized access to corporate networks and steal information or cause disruption.

To do that I would need to know their internal networks. To get this, I would need to perform some reconnaissance work and to do that work; I need to be monitoring the correct network traffic.

To do that I need access to a machine in that network. To get access to a machine in the network, I need to get my equipment connected to that same network (the IT request I made to obtain a port).

133

That equipment would need to be capable of performing reconnaissance tasks (the printer is the equipment), that equipment if it were the printer, would mean I hacked the printer and wrote a special driver. Hence, it behaved as a printer to everyone else but also could perform special operations.

That printer would need a way for me to interact with it so I can extract information from it, then repurpose it to carry out the tasks.

Maybe we say that instead of connecting the printer, we just did that temporarily but now we can switch it out for my laptop as an example, removing the need to create special printer drivers. Then I would need to bypass their logging that shows this unplug, re-plug event.

Assuming I did all of this correctly, I would then need to use some malware to elevate my privileges or steal a password, so I can gain access to these deeper networks, again avoiding all the corporate network monitoring systems.

Lastly, I would need to bypass their Firewalls and set up a way for data extraction to complete the work.

Of course, this doesn't cover any motive on my part, but it's not the point of the story. The critical factor is there is a fair amount of complexity to get to the premise.

Suppose there was a failure on the clients part at any of those steps, such as not having any monitoring, having networks connected directly, or not requiring me to have elevated rights. They have no firewalls, and someone gives me their username and password, for example.

In that case, they have much more significant concerns than a printer connected to an edge network via a port allowed and recorded by their IT teams. The diagram below is a typical setup only but should paint a picture of the story I described.

The unauthorized PC would be my laptop, and I would need to connect to one of the networks shown. Typically you expect to have to jump through a few hoops including firewalls, so I added those to the diagram as an example. I would expect a corporate network to be significantly more complex and secure than what is in the network topology diagram.

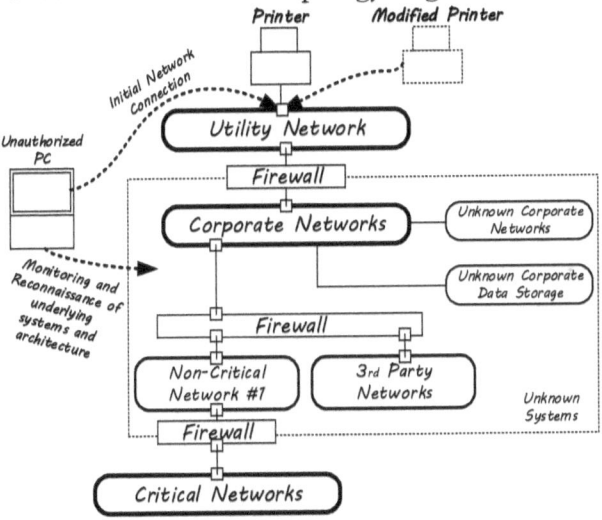

FIGURE 16 – TYPICAL HIGH-LEVEL NETWORK TOPOLOGY

The network architecture is an example of looking at a situation at the point of interaction, the connection of a printer and stripping it layer by layer to the fundamental part, which is stealing data by getting unauthorized access to a corporate network.

From the top premise, you check if it's true. Then you remove a layer of complexity, such as me gaining access to a port with a request.

Now what?

Then I removed another complexity. I had elevated privileges, another layer – I gained the network knowledge until the point where I could get to the corporate data.

While I used the method to break down a complicated situation for purposes of explaining myself, the process is the same for problem-solving.

Blood Sweat and Tears

The Strip step does not only refer to a layer model; it can also mean removing a variable from your equation.

When I was doing a job at a site in Queensland, Australia, one of the several tasks I had to do was upgrading the Random Access Memory (RAM) chips in a Thin Client (Basically a P.C.). A straightforward job.

I had the client on a bench, put on an antistatic wrist strap and got to work. I opened up the chassis, and there was a whole bunch of red dust—thick dust from a machine that had been running 24/7, in the middle-of-nowhere for over a decade.

I took a compressed can of air and sprayed it through the machine, creating what felt like a mini dust storm. I removed the old chips and got ready to put in the new ones.

The ram chip is a thin rectangle shaped Printed Circuit Board (PCB) and to install you simply align in the right direction and apply pressure to the top and bottom. For some reason it wasn't going in as smoothly as expected so I added a little extra force.

It struggled for a moment, but then it clicked in.

Phew, I was glad these were the right ones because there was no getting new ones for a few days as it would have to be shipped and delivered.

I finished the job, and I was about to close up the chassis when I noticed a drop of blood on my finger. I didn't feel it at the time, but I must have cut myself on the corner of the ram. I checked the chip, and yeah, it had a drop of blood on it. I wiped it down.

"Was this going to damage anything? Nah, no way", I thought.

So I closed everything and set it back up to its original location.

Now, I'm sure you can guess what happened next, but we had issues with the client after starting it back up.

It didn't start properly.

We began troubleshooting, and after a couple of hours with no luck, we continued the next day, and the thought my blood was causing an issue crept into my mind.

I couldn't shake it as a possibility in the back of my head. So I decided to "strip". This thin client ram upgrade was a 10-minute task maximum on our schedule, so the pressure was creeping up, and the annoyance of a menial task continuing was causing figurative tears. I decided to switch the ram chip out for the old one.

Did it fix the problem? Nope.

What I did achieve was eliminating an entire line of possibility. An issue with the physical chips, because as we mentioned in earlier steps, we consider if we made any changes? I also confirmed the new chips worked by plugging them into a completely different PC to isolate the RAM chips as a failure mode.

We look at the points of change and see if it made any difference (thinking back to Step 3, Eyes).

Using this reference point, it changed our question (Step 1) – was this machine having issues before the changes? To which the answer was yes. So we ended up re-deploying the operating system (Step 2) after I replaced the chip once again.

Problem solved.

How to Rapid Strip Piece-by-Piece

We have likely invested a fair amount of effort on a problem if we are at this step. That doesn't mean it is a time to slow down; in fact, it is the best time to speed up.

Again, not to rush, but to approach this step as if you were getting exercise repetitions in working out at the gym to gain muscles.

Problems solved at this stage are complex in nature. You usually are trying to find better questions, or you cannot reliably reproduce the problem, therefore having to deal with a spurious type event to occur.

Suppose you have a limited time window to access a system or an area to do testing or investigation.

In that case, you usually want to do more than one test, because if you check the results and it turns out you still have the problem, or you have no extra clues.

Then you would have to coordinate the various stakeholders to give you access once again.

This process takes a significant amount of time and leads to clients and managers asking questions about what exactly you are doing (or not doing).

Even if you expect the first thing you're going to try to work, ask yourself, "what if this doesn't work?"

"What would I try next? What if the results turn one way or the other?"

Make steps you would try in each decision branch and be prepared to do them all regardless of the result.

The advantage of approaching the problem this way is if no individual test turns up any clues, you might be able to see a trend across multiple tests, or weird behaviours or timings you notice in seven of the nine variations of your tests.

If you took a single COVID-19 test and the result was positive, you could conclude you are positive.

If you did three other tests that all came back with negative results, now what do you think?

STEP 7.
STRIP

The definitive version of this is to isolate your situation by removing a variable or stripping a layer of complexity. Run your tests, and then take away another layer of complexity/variable and perform the next batch of tests –before the analysis.

Solving Application Startup Issues with 10-minute test cycles – Part 2

Part 1 of this story occurred in Step 4, check yourself and just to recap it briefly before continuing here is, I was demonstrating an application as part of a FAT, and after closing the application, it wouldn't reopen.

I ended up doing some initial checks and couldn't fix it and finally settled on rebooting the server. The application ran again, and we completed the demonstration successfully.

Solving the root cause could now happen outside of the higher-pressure demonstration environment and work without worrying about causing unnecessary concern to any clients or management watching over your shoulder.

So what do we know now?

We know the application worked after a reboot of the system and it did not open again the second time without another reboot.

We also know we have run through several problem-solving steps up until this point so we will use that to create some rapid-fire tests.

I wrote down all of the points of interest, making sure I had eyes from Step 3 all ready to be a part of the testing.

I created columns for each, one for running processes, memory usage (ram), system event logs, application error logs, network traffic.

One column recording the time (in seconds) for how long it took the application to open and receiving the error.

I had another column for the timestamp of each test and finally if the test was successful or not.

In total, eight columns or data points per test with a couple extra for the test number and the test description.

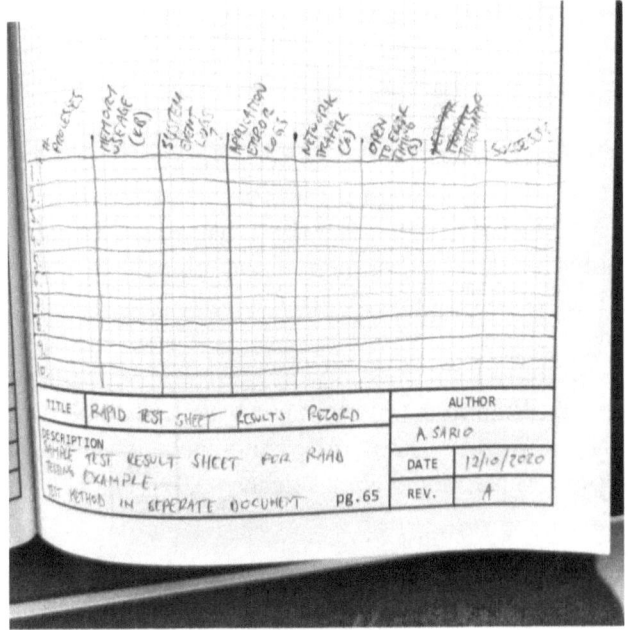

FIGURE 17 - RAPID TESTING SHEET EXAMPLE

STEP 7.
STRIP

The above image shows how the test sheet is drawn out in an engineering notebook, ready to write in the results.

Notice that I omitted the name or the test methodology from this sheet. Each test is a row which isolates a point of the system. I pre-wrote about ten various tests I would try.

1. Cold Start
2. Application closed using [X]
3. Application closed with the kill process
4. Application closed with Alt+F4
5. Run as administrator
6. Remote desktop into the system
7. Reinstall application
8. Increase configuration timeouts by 5%
9. Increase configuration timeouts by 200%
10. Close application, logout from Windows and log back in

This testing sheet came out to about 50 tests I tried. After running through those with all failures except for test 1, the cold start, rather than analyzing after each test (only spend a brief moment if something super obvious shouts at you) I moved onto the next.

After the ten tests, I made a variation and tried again. The application in this scenario wasn't a simple one. It integrated with three other servers, each with their applications running alongside it.

About one-third of the tests required a reboot, which we mentioned in step 4, took about ten minutes.

At the 50 test mark, my working window closed as the system had to be used by other colleagues.

I used this combination of information to come up with the root cause. None of the tests I tried on its own fixed the problem, but all the data points allowed me to find a pattern in particular combinations of tests.

So in this step, strip, you may not find a solution by removing one variable, you are merely proving something works, or behaves expectedly while in a simpler state.

Repeating and combining results from these various, more straightforward states can reveal the truth about the situation or the nature of a problem.

In summary, create a matrix of data points to collect as columns, based on Step 3, eyes, and Step 6 RTFM. For your tests, it will be using actions from Step 2 and 4, and you make variations of that to fill out all your Step 7 tests.

This process ensures your testing is equal, and that you monitored the same metrics for each test so patterns can emerge.

You learn very quickly that breaking things down to their fundamental truths will help you discover the root cause of your problems more often than not.

STEP 7.
STRIP

Your Checklist for this step

Strip

☐ Have I removed variables and checked result?

☐ Do I know for sure 1 = 1?

☐ What are the universal truths of this situation?

☐ Prepare a list of tests and perform rapid Strip
testing

☐ Have I reduced the problem to something so
simple it works

Removing variables and complexity helps us prove
what we think we know is true. When we say, "True" it
means, "true".

If we simplify, we may notice something to allow you
to go to a previous step and would signify progressing to
your answer. You can also use a combination of results
from different simplified states to formulate a solution
to your problem.

However, if your situation defies the laws of physics
or your relevant domain, maybe Mother Nature is not
happy.

What about the environment?
Go to step 8 to find out what to do next.

STEP 8.
WHAT ABOUT THE
ENVIRONMENT?

The grass is always greener on the other side

We have got to the stage where we see issues with even the very basics of what we thought we knew.

Step 7 had already stripped the complexity down to proving simple universal truths, yet we still can't find answers. Where else can we turn?

In this step, you will learn where to look next and open your eyes. My mother always used to say, "Look with your eyes and not with your mouth".

Looking at one spot and not finding the problem then proclaiming you can't find something prevents you from really seeing.

STEP 8.
WHAT ABOUT THE ENVIRONMENT?

When we talk about the environment, we are not necessarily talking about nature. We are referring to our surroundings and removing our lens for a moment. In Step 3, we discussed having eyes on the situation, but let's look in the dark for a moment. When camping and using a flashlight, you can see what is in the light but nothing else.

The most prominent blind spot is right where the light ends, not farther into the darkness. The second you turn off the flashlight you can't see momentarily, but then you can see more than when you had the light switched on.

So, what's this got to do with anything?

The environment is our surroundings. So take a look around, what are the external factors that change in the area you are handling?

Here are some example questions to think about:
- Do I always check this problem during my work shift?

See if there is some relation to the factors of your work shift. Someone else working at the same time, someone not working on something else at the same time.

- Does this issue happen at a certain point of the day?

Can you think of any changing factors or variables happening at the same time?

- What's the weather like today?

I'm serious on this one, has it been a rainy day and there is more precipitation affecting instrumentation?

Maybe it's been hot, and the temperature is having some interaction with some materials.

Or perhaps it's been raining for an unusually long time, and it's trapped some mice and forced them to chew through some cabling.

Now if you're thinking to yourself, "well these seem such incredibly low chance of being a part of the problem", well, you're right!

If you have come this far, though, it is appropriate to consider these things.

When looking at a slice of time, you have data, but when looking at multiple cuts of time, considering trends, you can get another picture.

"What did this look like this time last year?"

Solving a Problem that hides when you look

A few years back, we were close to completing a job, and the system, for the most part, was running well for some time – no issues observed for three weeks.

The system in this context was a few operator terminals with access to a control system. This control system was not doing anything critical at this point but was in use by operators.

After leaving, we got reports saying the system was "playing up" with a few "weird behaviours". Naturally, as soon as we looked, we could not find evidence of this behaviour, nor could we repeat it.

We spent about a day looking at various things and stepping through our problem-solving steps. We came up with a couple of hypotheses and set up more logging to get eyes on the situation.

STEP 8.
WHAT ABOUT THE ENVIRONMENT?

After a fresh restart of the system and monitoring for a while longer we did not see any issues for a few more days, so we left it alone and asked to be reported to if something happened again.

A week later – We got another report the issue had occurred.

We got into the system as soon as we could, and sure enough, it was all working fine. By this point, we are starting to get quite deep into our problem-solving process, but it wasn't until considering Step 8, "what about the environment?" we were getting closer to our answer.

So what are some of the environmental factors related to this situation we can consider?

Our focus here is Step 8 related questions, so imagine you have done the other steps, you investigated the "playing up" and "weird behaviours" reported. But aside from those diagnostic type pieces of information, the scenario had a few key components:

- The system had no issues for three weeks
- This problem occurred after we left
- The problem occurred after a week
- When we looked, we couldn't see the problem

By this step, you have tried obvious things, got extra logging, checked some fundamentals, looked up these weird behaviours, checked manuals and tried to repeat errors stripping back the system to some basics.

Now put your lens on the statements above, and this gives us a couple of new environment type questions:

1. Is the problem related to us logging into the system – try using a different user?
2. Are there any other things that run weekly?
3. What happened when we left?

Addressing environment question one: this is easily checked but gives us something to consider. We tested this, and it wasn't any different.

Environment question two now becomes interesting because we can check tasks scheduled to run weekly.

The strange part is we had the system running for three weeks so even if a weekly task runs, why didn't it have an impact in the three weeks prior? It became something to keep in mind.

Environment question three: What happened when we left? What actions do we take before leaving?

As it turns out, we enabled automatic backups which triggered weekly. The automatic backup could impact the running system, but if this were the root cause, we would have experienced the problem while we were there.

Again, Why didn't we see the problem when we were there for multiple weeks?

The difference was the total number of systems. While we were testing on-site, we checked each system backup individually, and it didn't impact the system.

But there were several system backups all triggering backups at the same time while the system was in operator use (more use than the engineering actions – think how much you use your car as a user versus how long a mechanic uses your vehicle).

This combination was what caused the issue to occur.

Not a perfect example, but it gives you an idea of the places you need to start to go when you are investigating a tricky problem.

It's a little clearer in the context of a system, but as an exercise, you may be able to recall a situation where the root cause was something about the surroundings of an issue.

Relationship Issues

At the beginning of this book, there is a quote by Eberhardt Rechtin on Systems Architecture. He explains a system is something you cannot produce with just a set of elements on their own. The key lies in the relationships between the parts.

How does this relate to our problem-solving ventures?

Not all problems apply to Systems Engineering although we are drawing from it because of the issues we are facing.

The "Spectrum of Problems" defined at the beginning of this book states that as problems become more complicated, the more inputs they have, which means more variables or moving parts combining to form part of your equation.

Your problem does not exist with a single input alone but the entire situation, similar to a system.

The key to these systems lies in their relationships. The "environment" talked about in this step may mean the data communications transmitting between the relationships.

It could also refer to a materials relationship with its melting point or the relationship of two objects leaning on each other.

We can't just focus on individual components of the problem we have to consider them in combination. We need to solve our relationship issues.

Assessing systematically naturally tends to the next question, what tools or techniques do we have for solving relationship issues?

One Drawing can solve 1,000 problems

The single best way to deal with these situations is the use of a drawing—the types of drawings that model the problem.

You can do this for anything, there are several standard drawing types used to solve specific problems so I can't cover everything here, but at Step 8 this is where you bring these tools out.

One thing every problem can use is just logically writing down a premise, putting it in a box, writing down another and then drawing an arrow starting from one to the next.

STEP 8.
WHAT ABOUT THE ENVIRONMENT?

Each premise could cause the next to occur, or list the events that happened in order. If you repeat this, you should visually see your chain of events.

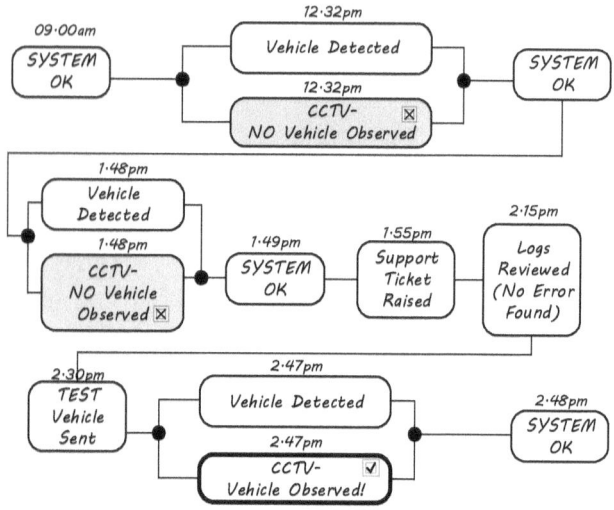

FIGURE 18 - CHAIN OF EVENTS

Now, in the chain of events drawing, you can change a variable and see where this affects every other link across your chain of events.

They say, "A picture says 1,000 words".
Well, I say, "one drawing can solve 1,000 problems".

My recommendations are swim lane diagrams (not only used for software or systems), architecture diagrams and cause and effect diagrams (fishbone).

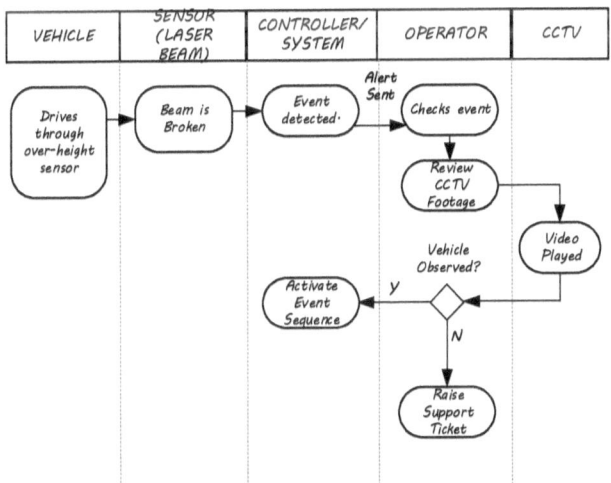

FIGURE 19 - SWIM LANE DIAGRAM

The Swim Lane diagram is powerful because you can quickly get the significant components down onto the page and separately focus on the events and interactions between these components.

You set out the major "things" into columns and then draw the process that occurs leading up to the problem. Now you can focus on which entities you need to account for when diagnosing the problem.

You can build upon the thought experiment of Step 3, where you take a journey of the process itself, or the energy transferring between different areas.

The swim lane example shows a vehicle monitoring process, but the power generation is also a great use-case to use in the swim lane diagram. Try it yourself. Go back to that electrical distribution and see if you could draw out the interaction.

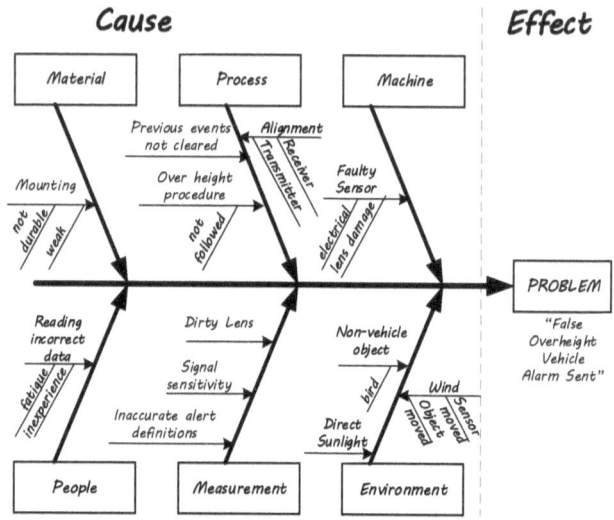

FIGURE 20 - FISH BONE CAUSE-AND-EFFECT DIAGRAM

When utilizing the fishbone diagram, you can almost think of it as reverse-engineering the situation. You separate the effects or symptoms you are having—one fishbone diagram for each symptom.

Then for that symptom, you fill in the possible causes across some key categories.

This tool is powerful because there could be several contributing factors to a fault, and you can understand which one to resolve first.

For completeness, I will address here the fact some people may say you should start with a fishbone diagram or our engineering processes should take care of these types of problems earlier and not even be a problem.

Ideally, yes, I agree. Established engineering processes solve problems - that's half the point of engineering – you must learn these for your field and use them effectively.

The way to think about the 10+1 method is the first chunk of steps can be done quite rapidly and solve many things. If every time a small issue came up you built entire system diagrams for analysis, you would spend a significant amount of time hypothesizing about something which may or may not solve your problem. Meanwhile, three or four issues could have been dealt with and moved on.

We want to reduce the mean-time to solve problems.

Now you might say, if you quick fix a bunch of things repeatedly it might be because of a bigger problem as the root cause – and you would be correct, an equivalent trend shows some other root cause.

But let's look at this logically.

If it's true that there is a deeper root cause, then you have just eliminated a bunch of underlying problems from being the root cause and arrived at the same conclusion.

In one situation we have only hypothesized.

In the other we have hypothesized AND eliminated several portions of the cause and effect chain, giving us both action and data.

There's also a chance you have lowered your risk profile in the earlier steps.

The Ghost Car

In the transport and roads industry, there are several sensors and signs you are managing. In some locations, we need to make sure vehicles are under a certain height so that they don't hit any structures.

This system is called an Overheight Vehicle Detection system and is essentially a laser beam mounted at a certain height. When a vehicle tall enough passes it, it breaks the laser beam, and it sends a notification to the operators.

One time there was a problem where the system would detect an overheight vehicle even when there wasn't one.

We had a *ghost* car on our hands.

Utilizing the logs we had in the system we could see that at 12.32 pm and later at 1:48 pm the beam was broken, meaning an overheight vehicle was passing by. When checking the CCTV footage at this time – no car broke the beam at the time.

We confirmed that the footage and the logs were time-synchronized, so both systems were reporting correctly.

The pattern seemed random. The following day we had one instance of it occurring and the day after that six instances. There were five other systems like this in the field, but they did not have the same issue.

No ghost cars detected. Perhaps the ghost only haunts one system?

We sent out a special vehicle that had an extender pole that could break the beam to confirm that it was still functioning correctly. So the truck drove out, broke the laser beam at 2.47 pm and in the CCTV footage, at 2.47 pm, we could see the truck breaking the beam.

Again, no clues.

So what's left?

For this problem, we arrive at Step 8, and the illustrations included in this chapter (the chain of events, swim lane and fishbone) all use this story to complete their example.

Now that you know the story go back and look at some of those diagrams.

What starts to jump out at you?

I know, you may not be the expert in this problem, and of course, you only have the information I have provided, but you can still build a feeling about where the problem could be.

Observing the chain of events, it becomes clear there is no pattern in timestamps, we do, however, see a common point where we have an assumption. "Vehicle Detected".

From the Swim Lane Diagram, we can see the genesis of the breakdown is at "Beam is Broken" at the sensor/laser beam.

Looking at the next column "controller/system", it seems to be functioning due to the message moving to the next system and its ability to initiate the event sequence later on when we activated the alarm with the test vehicle.

Finally, by completing our fishbone diagram and listing the cause and effect branches, we can look at all the scenarios and pinpoint our potential issue.

We did end up solving the problem.
Did you figure it out?

WHAT ABOUT THE ENVIRONMENT?

The problem turned out to be the sensor was mounted on a different type of pole to the others. This pole was strong enough to hold it, but if a large enough car drove by at a high enough speed (possibly over the limit), it would cause the pole to move.

Not enough to see on the CCTV footage, but sufficient for the laser beam to become misaligned from the sensor and therefore break momentarily.

Two other factors were making it hard to solve. One, by the time the detection event occurred, the vehicle would have already passed. The CCTV footage was correct – there was no car breaking the beam at the time.

Two, it also depended on the speed of the vehicle and not just the size, so whether overheight or not, the event could occur.

Water Your Plants

I was on an industrial site where we had to solve why this particular set of equipment failed on a specific day. Some other devices in an outdoor cubicle in the field controlled this equipment.

The system was commissioned about 24 months earlier and had been running 24/7 up until the error. After resetting everything, it continued to run for another month before having the issue reoccur.

The company deployed us to investigate, and after many failed attempts at breaking the system to reproduce the problem, we had nothing. At about this step in the problem-solving process, we considered the temperature. We knew if the temperatures got too high, there could be an impact on this system as it was out in the field.

The temperatures for the day of failure showed it was a sweltering 40°C (104°F) day, meaning this would be a pretty simple conclusion, except for the fact that the year prior had record-setting days of 43°C (109°F) and 45°C (113°F) but the system did not fail during those periods.

One conclusion might be the equipment is older than before and might not be able to handle as high temperatures anymore.

Not a bad environment type answer, but that wasn't it, this equipment life expectancy was 10 to 15 years. Moreover, we ended up finding the root cause, and it was also in the environment, in a different way.

Here's the breakdown.

There was a separate project six months earlier, which included decommissioning and physical removal of some equipment.

This old equipment used to cast a shadow on our equipment cubicle and so even on the same or higher temperature days in the previous year, the shadow meant the sun was not directly on this equipment for as long.

How were we meant to know what the shadow cast was like six months before we got there?

Before getting the final fix in, which was installing a shade structure to provide an air gap for reduced heat transfer, it would take a few days to install, but there was another high-temperature day approaching. We needed something to prevent a fault from keeping us running.

The temporary fix?

They sprayed the cubicle with water. (Yes, it was rated IP66).

They watered their plant.

Your Checklist for this step

What about the Environment?

☐ What are the surrounding variables of my situation?

☐ Has timing, weather, the person, or the temperature got any impact?

☐ What are the relationships in my situation?

☐ Have I got a drawing to model the problem?

☐ Did any other projects take place recently that changed the circumstances?

The grass is greener on the other side until you get there and realize it's not great, either. Your job is to check the "greener grass", go to the environment and check it is true.

Account for surrounding circumstances, including other project influences and see if you can approach the problem with this in mind.

If executing this step does not put you in any better situation, who you 'gonna call?

STEP 9.
PHONE-A-FRIEND

This Step could be earlier

Have you asked a colleague or someone else who may have dealt with your problem or similar?

The first push back you may have at this step is, "why not ask someone earlier instead of wasting all this time?"

For the most part, this is a fair sentiment, and we did say this could be earlier; in fact, this *could* be your first step. The goal of this problem-solving process is for you, specifically, to reduce your average time to solve problems.

To more consistently come up with solutions yourself. If your first step is to ask someone else, you become dependent and never really develop these skills. –remembering what we covered in Step 5.

I will walk you through the concept of "logical fallacies", "induction vs. deduction", using syllogisms, and how to get the most of your colleagues feedback.

The things you remember the most have a feeling associated with it. If you tried to conjure up all sorts of memories in detail from one year ago, it might be difficult, but as soon as you smell a particular fragrance or food, you are instantly brought back to a vivid memory from ten years ago.

The feeling we attach to problem-solving in this situation is the grief or the pain you might face when being stuck on a problem. If you are stuck on something for days and it is very frustrating, you don't soon forget the solution.

By passively deferring your brainpower to someone else, you never experience the mental journey of navigating a problem requiring a significant amount of time to solve. You have nothing to associate the memory with and is like every other thing someone else solved for you – you forget.

When is it Okay to Ask for Help?
It is always okay to ask for help in general, at Step 1, in particular, we ask many questions, but if you are to be an expert problem solver, the best time is at this step right here—Step 9.

You've tried several other things and have gained useful information at each step about solving your problem, but still have no answers.

Now you are asking someone to provide a new perspective, rather than getting them to repeat a step you had the possibility of doing yourself.

You may be thinking, "Okay, if I'm just asking someone, what more is there to this step?" You ask someone, and they know the answer, or they don't. True. But you have to read the whole chapter to reach the gold.

We will cover three things:

- What should I ask?
  - o The whole situation to now?
  - o Do I include the data?
  - o Do I include test results?
  - o The original question?
- How should I ask?
  - o Conversation
  - o Email
  - o Message
  - o Phone call
- Develop a Syllogism
  - o Understand my deductions and inductions

These three things will give you access to teamwork based problem-solving.

What Should I ask?

The first thing is to understand you need to make it easy for your colleague to answer you. The same in school, when wanting to increase the chances of getting marks for questions you've responded to, make it evident by showing them where to give you the grade as per the marking criteria.

So how do I formulate this?

Good screenplay dialogue practices tell us, for people to understand a scene the most important factors to convey are intention and obstacle. Anything else feels unnaturally scripted or forced. I'm not asking you to write a play. We just want to tell someone we have a problem and ask how to fix it.

We can still use this intention and obstacle concept. I understand your time crunch by this point, but there is no point of rushing.

You save time if you do this properly rather than re-explaining your questions repeatedly. Start with your intention. State the intention from a user or process perspective.

Consider this scenario.

A user tried to use your product, and it's a product that comes back to the person after they throw it, like a boomerang. But the product did not come back as expected.

You take away the product, and as you do Steps 1 to 8 of this problem-solving process, you found some weird behaviour in one of the isolated tests you ran. Now you are at Step 9 phone a friend.

When reaching out to someone, do you:

a) Tell them "when a user threw the product it didn't come back"? Or;

b) Tell them "it had a weird behaviour in one of your tests"?

These are both valid statements and problems that occurred, which both have pros and cons telling one story or the other in isolation. Sentence a) doesn't add any of your biases to the information while sentence b) saves time as other tests occurred before finding this strange behaviour.

The answer?
Do Both.

STEP 9.
PHONE-A-FRIEND

Start with the intention and obstacle. Sentence a), followed by b) and then stop.

Please stop.

In the interest of being succinct, this is probably the best way. What's the shortest path between 2 points? A straight line. By providing these 2 points, the beginning, and where you are now, they can quickly draw a linear sequence in their head to figure out what happened.

They will follow up with questions that come to their mind knowing these two pieces rapidly and in as least complexity possible.

It is effortless to get caught up in tangents, this is normal as many engineers fall into this trap, so just using this small technique can get you moving in the right direction faster.

How Should I ask?
There's no hard rule here on what is right or wrong as it isn't the best way to frame this. I'll just briefly go through the options and their pros and cons, so you have some confidence what to use and when.

There are three key drivers to consider:
1. Urgency
2. Traceability
3. Repeatability

If this is urgent, this step should likely have happened earlier, but it may also have switched statuses in this time. In either case, again, what's the shortest path between 2 points? Whichever method gets you directly to the person the fastest. Usually a phone call or a conversation.

If it is a little less urgent and the phone call would be interrupting something else going on, a message can work too with a simple "hey, I need assistance with a failed product, when is a good time to call?"

If it is going to be a long message and you want to put detail, do it in an email instead. An email has one advantage over the rest, which is traceability.

You can forward it to someone else who can help if more suitable.

To address the repeatability, you may want to ask questions again, so don't be difficult in any medium that you reach out with, be polite, but direct.

Rambling and not getting to the point makes it challenging for someone to help you. Consider now, instead of just asking for help; you are giving the person helping you even more work to figure out what it is you need help with, in the first place.

One of our goals here is to help others logically help us. So let's borrow from another famous engineering figure, the Father of Logic itself, Aristotle from Ancient Greece some 2300+ years ago.

Developing a Syllogism

When conveying a problem and the history around it to someone else, you have several angles to approach this.

PHONE-A-FRIEND

In some sense, you are constructing a conclusion that something is not working and it's because of a set of reasons.

This statement is a conclusion you developed off logic based on the input information.

The way you counter an argument is by proving either that the logic is incorrect, or that the inputs are false.

How we will use this is by turning your problem scenario into a conclusion, and then using a logical argumentative approach to counter it.

By countering the conclusion that something isn't working, we can find the point where we should focus on without needing the details of the problem itself.

How does it work?

We formulate a structure so that we can focus on the input or the logic.

What's the structure?

1) Major Proposition (a general statement),
2) Minor Proposition (a statement regarding the particular case),
3) Conclusion.

For a syllogism to be valid, the conclusion must be true when both the major and minor propositions are true.

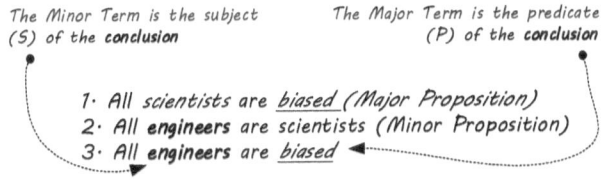

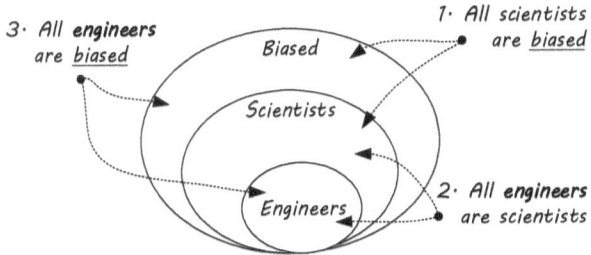

FIGURE 21 - STRUCTURE OF A SYLLOGISM

For the above, logically, we can conclude that all engineers are biased (because all scientists are biased), but not all biased people are engineers.

Imagine a scenario where a valve didn't open when an operator pushed the button to open the valve.

The root cause could be a failure of the signal or a mechanical failure. We can develop two syllogisms.

The Signal as a Root Cause Syllogism
Major proposition:
When the operator presses the button, it opens the valve.
Minor proposition:
The valve doesn't open without a valid signal.
Conclusion
The valve failed to open because of no valid signal.

Mechanical Failure as Root Cause Syllogism

Major proposition:

When the operator presses the button, it opens the valve.

Minor proposition:

The valve doesn't open when there is a mechanical failure.

Conclusion

The valve failed to open because of mechanical damage.

The diagram can help to get a visual understanding, but you don't need to draw one as long as you understand the core logic.

You will quickly find that some things appear to be so trivial in following their logic, but you will start to be surprised when a problem-solving process only follows one logical path to the fault.

You can look outside of your sphere and take multiple lenses at your problem and compare different root-causes you have hypothesized.

Categorizing them correctly and understanding the logical relationships will allow you to look for another angle, straight from your logic.

You can build the correct branches from your core logic and then utilize some of the diagrams from previous steps to map out your thinking.

Let's look at an example.

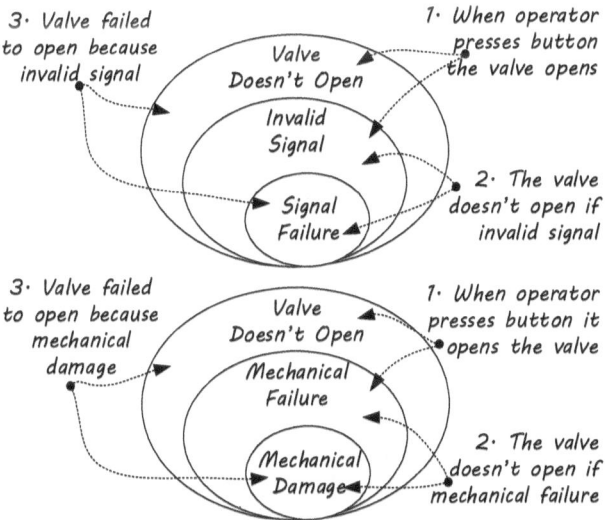

FIGURE 22 - PROCESS CONTROL SYLLOGISM COMPARISON

This syllogism might seem a strange way of looking at it, especially since it seems logical. That's because it is literally logical. We use this tool to help us solve problems by ensuring we haven't made a logical fallacy.

In the first line of reasoning, we see that the valve doesn't open due to an invalid signal. A signal failure could be one reason – it's logical, but there could be other root causes of an invalid signal.

Similarly, in the mechanical damage example, it could be a mechanical failure with one possible root cause being mechanical damage. But there are other types of mechanical failures.

What we have achieved here is knowing where to attack our logic. These are the 2 points of attack:
a) Is the logic solid?
b) Are the inputs true?

Don't Fall For a Logical Fallacy

How do we counter this argument, i.e. attempt to fix things here? We have to attack either the conclusion or look at the propositions and see if they are true.

Remembering that for the conclusion "the valve failed to open because of a mechanical failure", this statement would be false if either of the major or minor propositions were false.

Before delving further let me change the conclusion so that it is obvious the logic doesn't work. If the conclusion was instead "the valve doesn't open because the operator is sad" it doesn't hold up to the major or minor propositions.

So you would have a logical fallacy and end up spending all your time trying to figure out how to make the operator happy.

It is often the case you incorrectly use your logical conclusion to determine what to fix next. A false cause where you perceived a relationship (operator happiness) as the root cause of your problem. In this scenario, the operator might be unhappy because of the fault and not the other way around.

Let's switch focus to the minor proposition – the "valve doesn't open when there is a mechanical failure". Can we check if there is a mechanical failure?

Let's review the major proposition - "When the operator presses the button".

Did they *really* press it? How did they press it? Long, short, double-tap? Etc.

The only way you have your fault is if your major and minor propositions were both correct in the first place. Be sure to confirm these first.

The next question might now be, why did we develop a syllogism?

You want to be Right Something is wrong

Worse than having an unsolved problem is when you ask someone for help – and there is no problem. You are wrong about something being wrong. We want to be right about something being wrong, so we have something to fix.

If you say "the valve doesn't open", and they go over and open the valve, or it's already open, they can't help you with anything.

You want to have the logical reasoning to why you concluded something isn't working, so they can help you solve it.

If you have no argument formed, then this gives the person assisting a vague premise to work with and will essentially start again, or just repeat the steps you've already tried.

Black Screen of Death

At a remote site in South Australia, I was working with a small engineering team to commission a control room and the backend control system.

The control room had multiple operator interfaces, each with quad-screen setups.

STEP 9.
PHONE-A-FRIEND

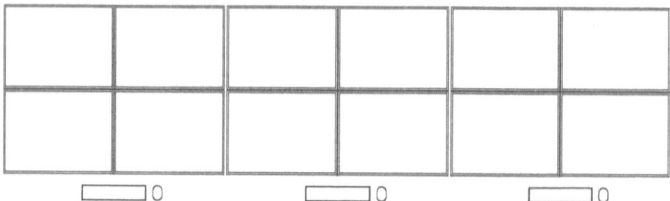

FIGURE 23 - MULTI SCREEN SETUP IN A CONTROL ROOM

A decent layout and a visually pleasing one, for sure. Each screen has sections of the process monitored with a range of alarm screens and controls.

Everything is nicely built up by our team and by the ninth day of us being there. The 10th day was the last of the trip and included a flight back home.

We allocated some time to clean and pack up our equipment, it was a super clean install (who said Engineers shouldn't do install) and we had practically handed over to the client.

This part of the project is where we can joke around a bit with the team and the client, have some laughs and exhale a little after a job well done.

Then, right before we loaded things into the car in the last hour of the day before saying goodbye, all those nicely setup monitor screens went black.

FIGURE 24 - BLACK SCREEN OF DEATH

Our hearts sank, the smiles disappearing before switching into problem-solving mode.

This predicament is the last thing an engineer wants to see in the last hour of a job, mentally preparing to fly home.

Usually, a screen goes black when it is powered off. However, there was power, and the monitors were indeed on. We checked error logs, but there was absolutely nothing. No warnings or errors and all the sounds worked, everything was "working".

If we described the problem, leaving out the screens displaying black detail to someone else, they would think we were wrong, and there was no fault.

We were wrong that something was wrong. Every single sign points to this conclusion.

This stage was undoubtedly the time to use the phone-a-friend card. We had support in another country and time zone where we could provide the support team with all the information to continue to work on the problem overnight while we slept.

Meanwhile, we were left to wait until the morning and pray for a good result. But before we get to the next morning, was there something we could consider from what we learned in this chapter?

Black Screen of Death Syllogism

Major proposition:
The screens go black when there is a fault
Minor proposition:
Our screens have gone black
Conclusion
There is a fault because the screens are black

Our logic here works so where do we attack next and check if it's true? Our minor proposition is true, we have observed the screens are black, but what about our major proposition "the screens go black when there is a fault"?

We know there is no fault, but is it true the screens display black only when they are in fault? No. It can't be.

What is another case when a screen displays black?

It's when the screens are told to display a black screen!

Maybe there is no error because it is correctly displaying a black screen.

Now a new set of questions can be made and should be investigated around where the system forms screen displays?

In the middle of the night with this set of thinking, we fed the information back to our support team while they investigated, but this is as far as we could go.

The next morning we were preparing for cancelling our flights as we drove to the customer site to check. We entered the control room sombre, and we looked.

The screens were live.

WHEW. "We're going home!" we all thought, but of course, there were some questions to answer. Our support team had found out one of the "profile configuration files" was empty.

The profile configuration contains instructions for the base layout of the screen.

Luckily, the support team could restore the files from a backup we took the day before.

Sheepishly, our client had admitted they "might have" been rummaging around in the administration folders.

We couldn't find a fault, but the answer was staring in our face. Think about it. The screens were correctly displaying the incorrect thing. We couldn't find a flaw because there was no fault – it was an undesired behaviour, sure, but this is why we check our logic when asking for help.

Smiles returned, and we could officially leave the site, make our flights on time and go home. I had time to reflect on the way back, and for the most part, there was a sigh of relief, while also thankful our company had overseas support.

Knowing someone else was working on the problem helped me detach from the immediate situation and think about the problem in this fashion.

When you phone-a-friend don't only transfer the problem to them, they are there to help you bear the load.

While you have the support isn't the time to relax, but a way for you to use the slight detachment from the problem to develop your logic, attack the reasoning itself or attack the propositions.

It gives you a moment to shift gears like they are your clutch in a clutch moment of desperation. Don't offload the responsibility to them; instead, solve the problem with a two-pronged attack.

Your Checklist for this step

Phone-A-Friend

☐   Have I done the problem-solving steps but still need help?

☐   Are there any other engineers who have dealt with a similar problem?

☐   What type of response do I need?

☐   Have I considered the urgency, traceability and repeatability of my situation?

☐   Is my logic sound?

☐   Can I attack my syllogism at the proposition?

A situation might seem dire, and as you run out of time, as a last resort, you shout for help – nobody answers.

You hope someone knows the answer to your problem, maybe they do.

If not?

Our hope turns to prayers in the next step. Pray.

STEP 10.
PRAY

Who will answer your call?

Never underestimate the power of prayer. I don't necessarily mean the spiritual side, although if it pleases you, don't hesitate to pray. If you're at this point into solving a problem and still have no answers, you just might need a miracle!

It certainly might feel like it. In all seriousness, what actions does one take when praying? It is likely some form of inner monologue articulating what you want addressing.

"Please, help me to solve this problem – please let this system run smoothly – PLEASE. LET. THIS WORK…"

I am not saying some other force will solve this for you, but I am saying to do this action, with one extra rule – actually detail out the problem. There is power in the act of explaining the situation to someone or even something else.

It might sound a little farfetched, but to our Software Engineering friends, this is a well-known phenomenon that can occur to the point where there is a decent meme.

The purpose of this step is to allow your subconscious mind to do some work. I'll show you how this is possible, and it does work.

Besides, you will also see how laying things out can help you answer problems all by themselves or at the very least create patterns previously hiding in plain sight.

The mind is quite a powerful thing, and we should be utilizing its full potential.

On the Spectrum of Problems, we are usually firmly secured in Type 2 or a Type 4 situation - above the maximum expected time to solve line or in other words, the threshold of stress.

So what is one way to actualize this step?

Quack, quack. The Rubber Duck Debugging Method

There's a method known as the rubber duck debugging method where a software engineer will have a physical toy duck sitting on their desk. And when they run into problems, they must explain the situation to the rubber duck. That's it.

It sounds insanely simple in principle, and it is. It is also what makes it intriguing.

The best part is this method works so often you may use this technique throughout your problem-solving process as early as Step 1.

But let's break this down for a moment. Why a rubber duck? A rubber duck isn't some placebo type mythical creature.

Still, it is so inanimate and innocent you have to simplify your problems and arguments to the capability of a rubber ducks understanding.

Since you have to simplify your argument, you are technically stripping down complexity and naturally trying to get to a root cause explanation.

This dialogue is robust because you have to answer yourself and come to some sort of reasoning mentally. Even if you don't solve the problem then and there you can get to a point where when you explain the sequence of events, you can see where you went wrong.

They say you truly understand something if you can explain it to a 5-year-old.

Using the Sequence of Events to Create a Pattern

Think about the sequence of events, from the initial observation, to triage, to all your problem-solving steps in between.

By doing this, you might find spots you missed, or a new combination of steps or a version of your tests you did in previous actions you haven't tried yet, so you jump back to any of the earlier stages. It's like noticing different details when you watch a movie for the second time.

PRAY

Imagine what you have is a ball of string representing your problem and situation. Imagine taking one loose end and then trying to unravel it, all the way until you have the string extended in a straight line as one continuous piece.

You can see how long this string is. When it was in a ball, it is impossible to tell how long the piece of string is.

The phrase "how long is a piece of string?" generally is used to say while you know there is a problem (the ball of string) until you extend it you don't know how long it is.

When solving a problem we know the scope, but when it comes to asking how long will it take to solve the problem, it is like asking how long is a piece of string: we will know how long it takes to solve the problem once it is solved.

We can't predict the length of the piece of string. But we can start to untangle the string by laying out the sequence of events.

You don't have to use a duck specifically. Sometimes this happens in the previous step as you try to explain things to someone else, or in Step 8 when you draw out the sequence of events.

Still, the difference is they respond and provide feedback and have some level of understanding you are trying to leverage.

Instead, find any inanimate object, I have a 3D Printed Minion from the movie Despicable Me, an Ironman and Yoda bust to serve alongside my rubber duck. Baymax from Big Hero 6 might also be a good option – get creative!

Although these are all just figurines, explaining to them individually or as a "crowd" for some reason makes me give them slightly different explanations. Or rather, I focus on other parts of a problem subconsciously.

To get the most out of your subconscious, you must visualize both the problem and the resolution. Try not just saying words and having dialogue but creating a mental image to give your subconscious something to hold on to and focus.

Where focus goes energy flows.

When you fill your mind with images of the problem-solving and potential resolutions, you might dream the answer in your sleep.

Some say they had nightmares about the problem, but it does happen. You know when you wake up in the middle of the night and just think "I've got it!" your eyes widen and something just clicked. You can make this happen more often than you think.

All we are doing is getting our conscious and our subconscious to team up and focus on the problem.

Designing an Artificial Intelligence Plugin for Control Systems – In My Dreams

I remember waking up from a dream and thinking "I need to write this down ASAP" before my memory fades into nothingness.

STEP 10.
PRAY

I jumped out of bed, scrambled to find my engineering notebook and a pen, and started furiously scribbling down the idea.

I didn't write words describing the idea, but I drew out the framework itself.

I had laid out where I would write the audio interpreter, how this would connect to some logic to interpret the commands, how those commands could convert into actionable interactions for the system.

What existing feature I could use and manipulate to generate the interface for these interactions, too.

I developed two things – the overall framework and the process for making it work.

It included a primary engine I would code, pre-programmed alarm triggers that would occur from the system, and also used inputs from an operators voice and the current system state.

We could manipulate one of the Human Machine Interfaces (HMI) to predict and present the most suitable screens to handle a given situation.

That's the work the subconscious mind got done, the conscious mind then went to work on to giving it a name, if there was a business case to present, thinking through the hurdles, and also validating the idea by walking through user experience scenarios.

Would this be useful?

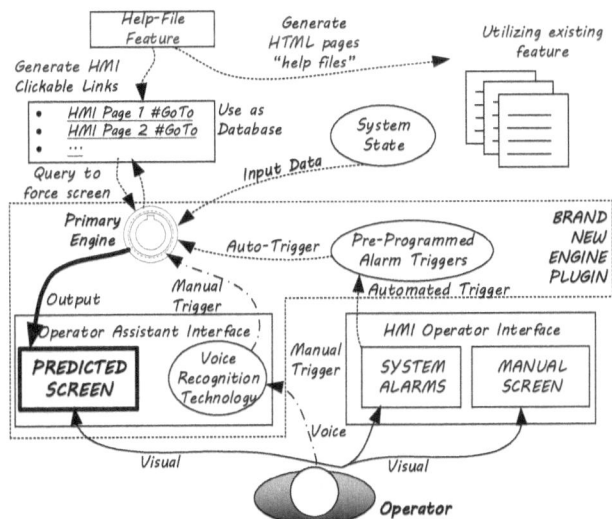

FIGURE 25 - AI OPERATOR ASSIST FRAMEWORK

I remember bringing the scribbled paper design into the office and a colleague asking, "Where's this from" to which I replied, "I dreamt it", I was met with disbelief.

Was this a one-off fluke?

It certainly isn't an everyday occurrence, but when it comes to designing a new solution or problem-solving a Type 4 problem, I commonly get answers or at the very least clues from my subconscious.

Now you might think, I don't remember my dreams, is this the only place the subconscious can give me answers?

It isn't.

Your army of rubber ducks may not be the only way to help walk your subconscious through a problem, sometimes once you have put all the work in you have to do something completely different.

Spinning on Chairs and Flying Toy Drones

I was working with a close colleague of mine on a brand new software solution, entirely developed in-house. As part of the process, we set ourselves up in an office with excellent collaboration space and limited distractions. Whiteboards, post-it notes all on the walls, printed off diagrams hanging, detailed documentation, network designs, system integrations and of course a bunch of code.

We were several weeks into development and on track for a progress demonstration deadline with the customer.

But it turned out something once thought as solved wasn't quite right.

Since we didn't solve the problem this far into the process, it had us caught between a rock and a hard place.

We practically had just this one thing left to solve, but it was a relatively large problem because it prevented other things to be demonstrable.

Although other areas were ready, they wouldn't appear as such without this one thing. If we were at 9/10 features prepared to demonstrate, the last part missing would only let us show 3/10 of them.

The client required us to have 10/10 demonstrated features to pass.

It bogged down the whole development process, and our progress came to a screeching halt.

We were now a week into this problem, and we did what any regular pair of engineers would do at this point of the process.

We maniacally laughed while spinning in circles on our chairs.

Okay, it wasn't as bleak as it sounds, but it felt like it at the time. This reaction wasn't out of panic, but it's similar to having one hundred things on your plate, and you decide for a moment to watch a movie instead.

Suddenly one of our interns interrupted us with a big grin on his face. A package just came in with this miniature drone he ordered online, about the size of two coins. It came in cute packaging and had a simple looking remote control.

In tandem, despite having this mountain of problems to solve, we became utterly distracted by this drone. You couldn't stop us from seeing how well this baby drone could fly!

What started off as just being amazed at this thing and taking turns just trying to control it, very quickly turned into a competition of who could fly it through a course made up of water bottles to a landing pad the size of a post-it note. Who could clear the course and land it in the shortest time?

To the untrained eye, we were playing games at work. Any passerby would see us flying around this mini drone, laughing and having a good time. "That doesn't sound like work" some would say.

After a solid half-hour of piloting and a major crash, we packed it up.

"What were we working on again?"

After catching ourselves up, all of a sudden, we could describe the problem in a way that leads to a solution.

It became clear as day, and we had a fix within the hour.

Aristotle Pacing in the Lyceum

When Aristotle was teaching in the Lyceum, he founded a school of philosophy called the Peripatetic. He designed it specifically to teach while walking along with his students.

This technique wasn't unique to Aristotle; other great famous thinkers of the past did this, too.

Why did he do this?

Basically to help him talk and think. Science has supported this concept with a Stanford study showing an increase in creative thinking when walking.

Even if you did your walking on a treadmill, there was still a boost.

The same study did not show the same results for more focused thinking, only creative thinking.

But shouldn't you be using focused thinking to solve the problem? Well, you have been doing focused thinking for quite some time, a solution at this point needs some creative thinking.

Disconnecting, getting into a flow state, taking a walk are all ways to help engage the early creative brain to start waking up.

Think about some activities you enjoy, the ones that genuinely grip you. You may risk getting distracted but just set yourself a timer.

You may be thinking at this point, "so, get a rubber duck, talk to it, pray, play games and sleep to find my answer?"

To that, I say, yes.

Your Checklist for this step

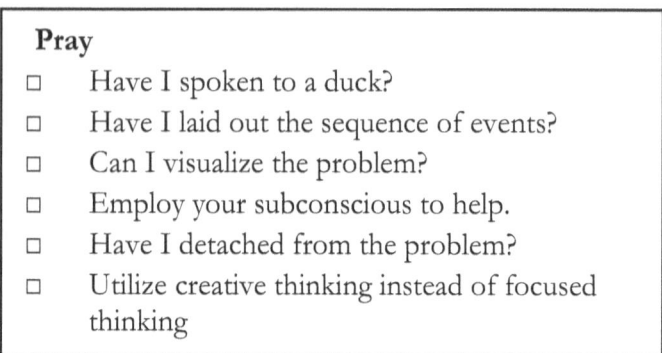

> **Pray**
> ☐ Have I spoken to a duck?
> ☐ Have I laid out the sequence of events?
> ☐ Can I visualize the problem?
> ☐ Employ your subconscious to help.
> ☐ Have I detached from the problem?
> ☐ Utilize creative thinking instead of focused thinking

Inspiration from this step should allow you a new angle to be able to return to a previous step. You can come up with things you weren't considering the first time through your problem-solving process.

Although reading through this book may seem like it is a long process, it isn't necessarily. If you go to each chapter, go to the last page and take the checklist for each step, and mentally tick them off, you will find you can jog through this process relatively quickly. I've also summarized them in the "cheat sheets" section for you.

You will start to remember each one of these and instantly come up with a bunch of things to do next. I designed the framework so you can jump around between steps and use the results of each to iterate on your progress.

You will get better at identifying when you can jump to another step, and you will soon start doing steps in tandem, gaining even more efficiency in how you arrive at conclusions.

I have always figured out most things with these ten steps, and now that you have this knowledge equipped, it should translate to you first improving your problem-solving skills and second building your reputation as someone who can reliably make problems go away.

Your skills develop as you do.

But, is this all? What if you're still stuck on a problem? Is there anything else we can do?

Now you have read through and understood the ten steps there is only one possible step left you could try. In addition to the ten steps, there is one more to add.

STEP 11.
THE SECRET STEP

10+1 – Clutching at Straws

Fix what the problem is not.

Congratulations on learning the problem-solving concepts in this book so far and making your way to expert engineering status.

If you're here though, it does mean you are dealing with a problem well over the expected time to solve line in the Spectrum of Problems outlined in the introduction.

There is desperation where it feels like you've tried everything and you are clutching at straws. You don't have anything left, no ideas to try, no clues and no answers.

Even after all ten steps above, including repeating a few of them, you are still at a loss. You started with The Question and questioned your inputs, and then you moved on to all The Obvious fixes you knew and executed all of them. You placed the correct Eyes on the situation, with all possible logs you can think of, checked monitoring tools and sensors – in all the places where change takes place.

You confirmed the fundamentals, dealing with Layer 1, reset things and repeated things with no results. You used doctor google effectively and utilized online resources, you know them all. You have also referred to the manuals deploying the RTFM protocol, finding datasheets, manufacturer's information, etc. You wrote your own manual and had tests in place ready to go.

You stripped the situation down of complexity, starting with the top layers of dependency down to more specific conditions. You isolated variables in your case, but you didn't find anything.

You did not find clues in making something work in the simplest form, aligning with scientific truths. You set up rapid-fire testing of combinations of things you tried in Step 2 and 4 utilizing sufficient checks from Step 3.

You considered the environment, the situation, timings, weather, circumstances and other projects potentially having an impact.

You've asked the right people, subject matter experts, relative stakeholders and still haven't got anything substantiative. Or you are the most expert person on the subject, so if there's anyone who knows the answer, it's you. No one else can help.

You prayed to the engineering gods to help you with a miracle, you utilized the rubber duck engineering method, explained and understood everything about your situation intimately. You also managed to hand over some of the workloads to your subconscious mind to try and get some answers.

You reviewed your questions and performed variations of tests based on the new questions, and you've collected data from everything above.

Identifying You Have Tried Everything and Need a Secret Step

The point of summarizing all the steps covered so far is to ensure you have actually gone through all of them and realize you have tried many, many things.

You are not here in vain; you have spent considerable time and responsibly and expertly looked at the situation.

Let's say you are helping a colleague work on their problem.

"What have you tried so far?" you ask.

They describe the situation so far, similar to having completed the ten steps. Go back and forth, confirming they've tried different things until you believe they have tried everything that *should* work.

People working through this step will usually find the solution to their problem was actually within the first ten steps.

You will only know this in retrospect, but it is worth mentioning, so one gives the appropriate effort in the methodology. The more confident you are, you have tried everything, the more effective the options are in this step.

In either case, based on what your colleague has described, mentally check they have done everything.

The Loyalty of Unsolved Problems

I was working with an engineer from one of our direct competitors. In the electrical engineering world, this happens more often than you'd think.

Specific systems require a special redundancy which caters for having a dual system available if one fails, and also the system has to be from a completely different vendor—catering for any problems at a company level.

The vendor differentiation is a fundamental philosophy of protection systems. No common mode of failure even at a specific setting or configuration.

How often do you include an entire company going under as part of your contingency measures?

We had an interface setup between both systems for validation of outputs, and after about two hours of things going well in commissioning, the following eight hours we were dealing with a discrepancy. Both systems thought they were right, but the answers didn't match.

Now, you can imagine how easy it is to point the finger and blame the other party. After all, they are your competitor. For even more self-validation, everything on your side says you are correct. What more can you even check?

"I've done everything right, and I double-checked everything, the problem is on the other system".

THE SECRET STEP

The problem with this way of thinking is the person on the other side of the situation has used the same logic to arrive at the same conclusion as you.

Since I already know the fastest way to solve problems is to assume I know nothing, I humoured checks on my side, and I knew I was working with a competent engineer because he reciprocated.

Blame was not the goal; the overall solution for the client was the goal.

We were both starting to feel the pressure as it was a 10-hour testing window, and after 10 hours, we only had 20% progress to report.

Explaining this to management is not a comfortable conversation. We both stayed back an extra hour or so trying to make things work.

At this point, I had to call it.

"Hey, I think we should call it a day. We've tried everything we could and getting some rest will probably help."

He nodded but was a little hesitant, as he did not take his eyes off the screen.

"Don't worry." I said, "The problem will be waiting for you in the morning, I promise".

He laughed and agreed.

The next day we had our morning coffees, attended the safety toolbox meetings and provided brief updates on what we were working on during the shift. We went back to our workstations and guess what happened next.

We solved the problem in 10 minutes flat.

Now, this isn't the action of this step. The point is to give you an understanding that problems are loyal. They don't go away by themselves, and they will be there waiting for you when you get back.

So take a break, reset, heck, cheat on it and work on other problems, call them your side problems.

The original problem is so loyal it will still be waiting for you the next day. Let it go.

Be Leonardo da Vinci Level Observant

Da Vinci is another engineering legend who many in public may know as the artist who painted the Mona Lisa, the Last Supper or the Vitruvian man.

More relevant than those historical achievements was what he achieved in the realm of engineering.

His notebooks are a testament to his work. You get a glimpse of how his mind worked. If you go through his drawings and take the context of the era, there was no way to get the full picture of details he drew.

He had a great combination of attention to detail and the skills to express insightful observations in his art. So how did his drawings have such great detail? Well, it is where he paid attention.

There's a well-documented story about this. Imagine you are looking outside and you see a bird land. What do you think? "Look, a bird."

STEP 11.
THE SECRET STEP

Da Vinci rather, wondered how exactly it is the bird landed and if it did something to reduce its velocity. He noted the bird tucked its tail along with the last flapping of the wings before landing. Is this a coincidence?

How does this relate to the rest of the joints in the bird's body? What are the aerodynamics of all this working?

You might just calculate the force generated by the wings without considering the impact of the tail unless you noticed that little detail.

You see, when I mentioned before solving your problem likely lives within one of the earlier steps, it's because you weren't looking with enough detail.

Your standard of looking with detail is too low. Observing your situation in "da Vinci level" detail might give you a whole avenue you missed on your first run.

The Known versus the Unknown versus the Unknowable

In Science, we deal with the Known and the Unknown. The Known is what they teach and show in museums.

Whereas the Unknown is what scientists are looking for in an attempt to move them into the Known.

i.e. when gravity was undiscovered, it was unknown.

When the apple fell on Newton's head, it moved it into the known.

The unknowable is something that science doesn't answer. However, there's no exact scientific answer to define a question as unknowable (is there a prime mover? are we in a simulation? what is the meaning of life?), we can use this principle to think about moving problems from the unknown to the known.

In our engineering problems, we can think of the known problems with known solutions as the things we try in Step 2. The more you use this process, you learn to move everything unknown into the known.

An unknowable question might be "what was the designer of this thing thinking?"

When you get frustrated with a problem, you begin trying to think about the engineer, the human who made the decisions to design something a certain way.

The original designers real intent is unknowable. You can't test or find out. Now typically I would advise you to avoid this type of thinking.

However, there is a way we can utilize the unknowable, though. As a person in your field, you start to imagine if you designed a solution, what would it do and how would it do it?

You understand the several possibilities, and you build this into your logic.

I can typically find problems with existing solutions pretty quickly because I know the underlying principles. I extend this by trying to create my version of the solution, even if it is just a hacky version.

For example, when working in control systems, I have built my own mini control system at home. For working on networks, I have created a home network to solve particular problems.

For working with web servers, I host my own equipment and run a website. When I had a project with databases, I created a database for tracking home inventory.

You understand the intended solutions, and it helps you figure out how something might get designed poorly – because you considered doing it similarly hacky due to some constraint and perhaps you decided to stay away from taking any shortcuts.

The engineer might have had the same constraint and put it into the solution.

Now we have a means of utilizing the unknowable to help us push an unknown into the known.

Something from Nothing

Getting something for nothing is not a reality.

Think of a new building it's as if it came from nothing, but in fact, we just dug up the ground and materials from somewhere else, we went negative somewhere else.

I'm not asking you to create an answer from nothing, the same way you wouldn't consider the problem occurred from nothing.

Where is this ground we can go negative in so we can come up with solutions from "nothing"?

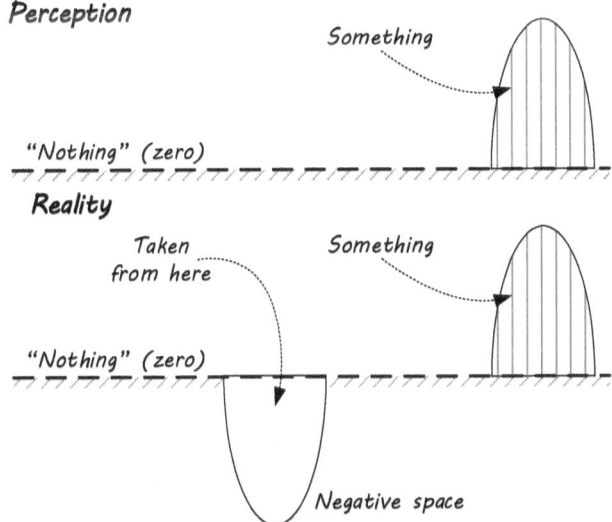

FIGURE 26 - SOMETHING FROM NOTHING

You need to put in hours developing the solutions to your solutions. You need to build your version of things so you can create a process of how to approach problems related to your situation. Then when things break, and it's due to a deficiency in a solution, you can almost predict it.

You can reverse engineer the situation and seemingly come up with a solution from "nothing".

Reverse Engineering
We have already tried reverse engineering our problem within the first ten steps, so how else can reverse engineering help us?

I was working on a proprietary software solution where they configured the entire operating system to optimize the software implementation. This configuration meant the designers coupled the software to the underlying infrastructure. The software solution utilized features of the operating system.

Due to this, the system requires experts or the original developers to handle any possible fixes as they have access to underlying dependencies and code to know where to look.

It can be challenging for engineers to understand why people build things with such dependencies and complexity.

Still, having made custom software myself to create an ad-hoc control system for my home, I knew there were certain conveniences and small advantages of utilizing the underlying system structure.

For example, managing files and storage within your application, or just letting the operating system do the job, and we merely interact with it?

Persistent storage of data needs to be addressed with an application because once your application ends, you lose the active data in memory. So you store it. Usually, you might put this in a database if it exists, or you could create it into a data file for storage on your operating system.

For the solution I built at home, I just encoded everything into an XML file and stored it as a text file in the project folder. This approach is not how you would do it in an enterprise-level solution, but there could be an instance.

The system problem we were having ended up relating to a configuration issue. Usually, this means combing through the connected database and checking if the data at the source is correct. Then reasoning up from there to see if either the process updating the database is at fault, or the system reading the value out of the database is at fault.

The issue was a missing parameter in the database addressing the configuration. If you checked the entire database of information and it doesn't exist where can you look?

I automatically knew the other place you could store information was in a configuration file. You could use this to keep the initial state the application should be in before even talking to the database.

Maybe it needs to know which database it should be in or if it is operating in a particular mode.

This solution was slightly more sophisticated than my home project, so they stored the file in hidden folders. We found where the configuration was wrong and managed to fix it so we could continue the job manually.

The root problem to solve though is:

1) The installation process getting this configuration wrong and;

2) How do we get informed the configuration has been changed or has an error?

Both jobs we could offload to development as feedback for improvement, but we could get moving without waiting for remote support to come back to us with answers after a few days.

Reverse engineering a different working solution can give you insights to the problem you are facing even though it seems unrelated.

The Seemingly Unrelated

So far, this Step has covered that we take a break as the problems will wait, we need to raise our quality of observation, we use the unknowable by putting ourselves in the designers head. We become a designer ourself and build equivalent solutions, and finally, we reverse engineer other similar solutions.

We are clutching at straws in this step after all, what do we do now?

My recommendation is to try tests that are seemingly unrelated to the problem at all.

Pull a random cable, test with a random system off/on, try it three times in a row, and try it specifically after restarting or re-plugging something in. Wait an arbitrary amount of time between steps.

Were you having problems with a traffic light? Try to run the water system. CAD program struggling to mate components you designed? Try to take a screenshot, edit it in Microsoft Paint and send the result in an email.

Now, not completely random but within some sort of reason. These tests are the ones when someone is having a problem, and you suggest it to them, they reply

"That shouldn't fix the problem, it shouldn't be related", and they are absolutely correct.

But here's the thing -- at this stage, they have already tried everything that SHOULD fix the problem (covered in the ten steps).

It's time to try the things that shouldn't.

This method works wonders for many reasons:

1. You are trying "anything" at this point.
2. Most of the time we may think we have Step 1 covered, but we don't.
3. Triggering correlations

The third item is the most important reason.

Triggering Correlations

Sometimes all you want to do is throw in new inputs to the system or problem you are solving to get clues or give new ways to approach earlier problem-solving steps.

There you have it. Problem-solve like a ninja.

Approach an extremely experienced person dealing with an issue and ask them what the problem is. As they describe all the things they've tried, throw in a random something they haven't.

And when they say, well it shouldn't fix it simply ask them

"If you've exhausted everything that *should* have worked, this is the time to try things that shouldn't".

STEP 11.
THE SECRET STEP

Either they will think of more tests they haven't considered to avoid doing your preposterous idea, or they try it and get a new clue to their problem.

Heck, at worst they confirm they do know something about the system.

Let's build on the preposterous idea for a moment. An engineer will always take a logical approach to things, so suggesting something ridiculous will automatically trigger them to disprove that it has any relation or would result in any success.

They build more links to the situation from a completely different vantage point.

Correlations don't always present themselves to you, so finding more and more correlations is useful because we know for sure we don't have causation.

Correlation does not equal causation, but if you gather enough of them, you might just find the cause.

Follow the Cluster F**k

There's an old saying for problem-solving along the lines of:

"Follow the cluster f**k".

Elegant, I know, but there may be no better way of phrasing it.

It's similar to crime shows, where ultimately, you "Follow the money".

For problem-solving, we are talking about where the most problems are, where they arise, and common perpetrators.

The problem you may be solving might be a part of a system with one issue, but if you look for where there are more issues related to one object, device, process, system or material you will eventually find the missing piece.

The difference between investigating a crime and investigating a problem is problems aren't usually malicious.

By following the chain of car crashes, you often find the path to the root cause.

Reporting the Unsolvable to Management and Clients

Dealing with difficult problems means there is already a big fire to put out and we don't want to be adding fuel to the fire.

Unfortunately, balancing both client and management expectations is another fire to put out. Typically the lead or senior engineers will deal with this but even if you are not the lead for the project you will be reporting to your senior who will report upwards to management.

The reason for including this is to remove a variable of stress and wasted brain cycles for reporting. The beauty of this process is you can cherry-pick from the completed steps and build a concise report.

It won't take much time, and it shows both the depth of the situation and your efforts, too.

It also gives confidence you are doing everything possible to solve it, which helps.

What information should you pick?

Start at Step 9, the information you collated to

phone a friend of the situation is already in a format ready to for consumption by others.

Next, if you are reporting technical findings for your senior engineers include all the tests and results from Step 6 and Step 7.

The statement before the attachments should address what you've accomplished so far and what is next.

"We tried 25 tests as shown in the list, so far we have found some clues but have not found the root cause, yet. The next steps will be to test three more things, and we will report from there."

In essence, you want to cover
1. What is the problem?
2. What is the current status?
3. What have we done?
4. What are our next steps?

After listing the next steps also include if there are any roadblocks you need them to remove. Think along the lines of gaining permissions, upward reporting, or providing a tool or resource, even if it is minuscule.

You should do this so they can help as much as they can, bringing them along with you to find the solution, as opposed to going against you.

18 Hour Critical Situation Turnaround

"Andrew, I know you just came back from the site a day ago, but we got a call from the client with a critical problem preventing operations coming online, none of the team on-site can fix this."

There was an urgent matter on-site, and I needed to go because a timing window was about to close that would lead to fines from the government to the client and then subsequently, passed on to us if we missed the deadline.

The issue wasn't just the financial impact, but there is also an associated reputational impact for all parties—a real situation when working in critical infrastructure.

Regardless of who's at fault or the current frustrations of the teams, ultimately, we were fighting the same battle.

Keep this in perspective when you are dealing with high profile cases and complex problems, so you don't introduce your own emotions into the situation. The site was remote and required driving 2-3 hours to get there.

After a short briefing, I was off to site, I dressed in full PPE, high visibility clothing before getting into the car because I knew I would be jumping right into the situation upon arrival.

When I got there, the first thing I did after safety checks (yes, your safety first even in a mission-critical emergency) was that - I listened.

STEP 11.
THE SECRET STEP

I started with the supervisors/management and then went straight to the closest to the situation, the operators.

I only got a high-level picture from supervisors, but I made sure to get and set their expectations.

"I understand the situation. I'm going to start straight away, are you happy for me to talk directly to the operators to get the details and set myself up in this area to get to work?"

He nodded.

"Do you need me to fill out any other forms or checks before getting started?"

"No, dive straight in, nothing is running at the moment."

I've got my boundaries, so now I could work freely. I had to move quickly, but not rush. By rushing, it looks as if that's how you got the problem in the first place, but dragging your feet would look bad too.

I approached the operator and needed to get his understanding of the situation, using open questions from Step 1. Take note: this is the most crucial time to avoid loaded questions or too many recall questions.

"It was working fine for the whole day until a few hours ago, I got these three alarms, and then the system tripped. The alarms are now gone." He corroborated.

I already had this report, so I asked if he had noticed anything else different at all, any small things.

He had a few more, so I noted these down and informed him I was going to set myself up, collect my logs, but before I take any significant actions, I'll let him know and keep him in the loop.

And that's what I did exactly. If you consider the whole situation, this was not the time for Step 2 the obvious, as we should have already tried this previously and it was successful for over 24 hours. No, this, was the time for Step 3, eyes. As I collected data, I began documenting. The initial report, the additional comments, the time and the fact I was collecting logs.

I spent quite some time analyzing this problem, and I added every piece of data and correlation I made to the document.

By doing this, I was essentially trying to build up a plan for tests in Step 6 and 7, and include the next steps for each outcome.

To get to a root cause and a workable solution to prevent a pretty disastrous project outcome from happening, I had to lean heavily on steps 6-11. I came up with some ideas using a combination of these steps and had my implementation document ready.

I circulated this to the client supervisors and project managers and then talked the operator.

"I've collected the logs and done some analysis, and I have a few tests I'm going to try. They shouldn't impact you, but please tell me if you notice anything strange. I also informed everyone else, and they were happy. All good for you?"

He didn't need to know the analysis. He needed to understand how the system may respond and if he needed to do anything. The same way a mechanic might inform a driver to tell them if they hear any weird noises while they do a test run. He had no objections and was happy to be kept in the loop.

I ran some of my tests with the results pointing to one of my theories. The theory was there must be a bug in one of the submodules which only our development team could resolve as it was not accessible by me.

This submodule shouldn't relate to the problem – but it was. Reaching this point, I had already tried everything that *should* have worked, and we learn in this step that it's the time to try things that shouldn't. I then managed to trigger a correlation with this module that contained the bug.

Thanks to my software background, I could help out development teams with detailed analysis. So as part of Step 9, phone a friend, I sent these findings and my conclusion to the support team overseas. I then proceeded to put in a temporary workaround.

It's the equivalent of someone being broken down in their car and pointing out to them they had a faulty part. You still have to deal with the problem of getting the car to a mechanic or somewhere to park and to get the driver home safe, too.

A Temporary and Long Term Plan

Since I identified the root cause, I was able to come up with a plan to put the system into a state where it could operate, albeit with a couple of alarms and not an ideal state.

A Band-Aid fix, for sure. It wasn't as crude sounding, but with my strategy, the operators could get running while our development team writes us a patch.

Using the broken down car example, I managed to toe the car to the mechanic and drop off the driver to their residence.

I identified the risks and explained to the operator what might happen, but I was going to put it to management and the relevant stakeholders for a decision.

Now, if I had gone to the site and rushed to find a conclusion, testing a bunch of things and ended up with the same Band-Aid fix and told them at the end, I fixed it, this would not go down well.

The client expectations and the stakeholder expectations will be all over the place, the direction, due diligence, documentation would all be missing, and you would be adding more problems to your original problem.

Not to mention, if you had any other follow up issues, you have nothing to build up from, and for your internal support teams, you would not have a document for them to read and catch up on the situation. It's unprofessional.

I let the system run for a couple of hours, monitoring its stability and then handed over to my team on-site, I emailed all relevant people and headed home.

By the time I got home, I was exhausted, and I was a little worried the fix in place was less than ideal, unsure of what the client would think. It was a pretty intense 18-hour turnaround, but I managed to sleep.

So naturally, you may wonder why I am telling you this.

An Unexpected Outcome

Aside from reiterating that the value of the +1 step is to try what shouldn't work, there was also an unexpected outcome.

The next day my management told me the client was pleased with our quick response. I got a special commendation from their top-level management as there were meetings on their side to report on the situation.

In my future performance reviews, a quote from a client's upper management directly commending my performance bodes well for justifying some sort of incentive – and it did.

This story isn't to show off that outcome, nor was it my intention for doing my job. I wanted to show you how one goal of this book is to improve your problem-solving skills, and the other part is how you approach your process – what type of engineer you are.

An engineer is a professional with responsibilities for both, solving problems and how you act, treat and work with others.

With this particular problem, I had to use all the steps I had learned, and having the +1 Step allowed me to try what shouldn't work to find a correlation which leads to finding the root cause.

The communication and documentation approach allowed me to manage expectations and enable other engineers to assist, too. As a team, we could achieve a great result, and as an individual, I could convert my problem-solving skills directly into career outcomes.

Hierarchy of Engineering Controls applied to Problem-solving

There is still one more concept you can use to help you come up with solutions. Another tool for your kit, building on the ideas you now know.

I use this outside of the engineering world and

think it has such remarkable parallels that I did a complete episode on it on the Engineering IRL podcast.

In your Engineering course, you would have learned about the Hierarchy of Controls. It is used for risk management and is particularly useful for hazards and safety.

1. Eliminate
2. Substitute
3. Engineering Controls
4. Administrative Controls
5. PPE

Because number 1 is the most effective for removing the hazard but also the least practicable, you can't always do this. So you try and do the next best thing and move down the list.

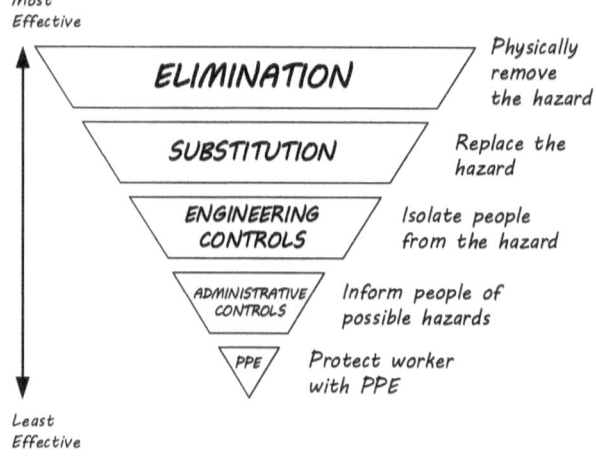

FIGURE 27 - HIERARCHY OF ENGINEERING CONTROLS

During my career, I have found some parallels that extend the hierarchy of controls concept beyond the realm of hazard and safety control.

If you frame the problems you're fixing as a hazard; you can assess your solutions against the above categories.

When things aren't working, you should try to categorize the solutions you have been attempting against the categories of the hierarchy of controls.

What tends to happen is you can see if all of your answers are actually of one type of control, forcing you to come up with solutions in the higher tiers. It is quite a powerful concept for returning to Step 1.

For example, imagine you are driving in your car, and you see a speed limit sign showing 60.

If you decide to exceed this and travel at 70, how effective do you think a second sign stating the limit is 60 would be?

What about a spill, if you place a sign where water is on the floor and people keep slipping over, how effective do you think another sign would be? Perhaps it's time to use a physical barrier, moving us up the hierarchy of controls.

Hierarchy of Controls for a Technical Problem

I was having a problem with a Microsoft Access Database, handling multi-user management. This database was a preexisting tool we had always used. There were several fixes I tried in the past, and a few more I could still try.

I had administrative controls as there were instructions on how to use the forms with comments included in the user interface.

I had to contend with the many scripts in this thing. I could write several pieces of code to address some of the issues, but upon assessment against the hierarchy of controls, these were all administration and engineering controls. It was time to move up the hierarchy of controls.

Could I find a solution by substitution or elimination?

Is there an alternate method for multi-user management? Alternate software? These suggestions might be too expensive at this point. What about eliminating MS Access and using a different application with built-in support for the feature?

You begin to ask questions which help you build solutions for the problem in a more controlled manner.

Hierarchy of Controls for a Real-Life Problem

What about real life? If you're having back pain and poor posture, there is a method out there to put reminders to straighten your back or have someone who sees you often to tap your back every time they see you slouching to remind you.

I found a product out there which attaches to your back and monitors if you are slouching. If it detects that you are slouching, it vibrates to notify you to straighten your back. I thought this was excellent, instead of relying on others or random objects to remind me, I have a device to monitor all the time.

It's a pretty good solution, but the problem is, it didn't fix my problem.

I looked at the hierarchy of controls, and at the end of the day, it was still an administrative control and even using all three solutions combined (the post-it notes, a friend and the monitoring device), they were all administrative controls.

So what's next up on the hierarchy of controls?
An Engineering Control.

Perhaps I need a back support brace – more of an engineering control because the slouch is a habit of over 25 years, it needs something more practical.
Had it been earlier on, perhaps administrative controls would suffice.

The example is just one way you can start to think about the solutions in both your career and your life.

The Secret
The secret to problem-solving is now out of the bag. It is a little unorthodox, and it isn't a straightforward machine that simply churns out repetitive work.

Now that you understand the tactics, you may be considering the importance of all of this. The truth is that your problem-solving skills are one of the least commoditized assets you can bring as an engineer.

STEP 11.
THE SECRET STEP

Anyone can follow an instruction set, learn to use a specific tool, write some code, draw an architecture, make plans, sign documents, make presentations and send a report. With AI coming into prominence, some of these number-crunching analytic type tasks also become a commodity, not even requiring a human.

So, stay ahead of the game, keep your unique problem-solving tools sharp and continue the art of engineering.

Your Checklist for this step

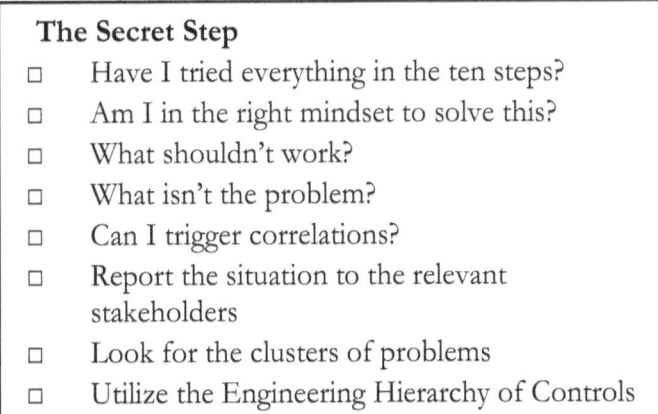

The Secret Step

☐ Have I tried everything in the ten steps?

☐ Am I in the right mindset to solve this?

☐ What shouldn't work?

☐ What isn't the problem?

☐ Can I trigger correlations?

☐ Report the situation to the relevant stakeholders

☐ Look for the clusters of problems

☐ Utilize the Engineering Hierarchy of Controls

Inspiration from this step should allow you a new angle to be able to return to a previous step. You can come up with things you weren't considering the first time through your problem-solving process.

Documenting professionally and approaching things with some direction will build your confidence. And now having the awareness on how you can translate some of these skills into career objectives will help you land more significant projects, responsibilities and roles.

You should find by running through these steps you may have already done them without even knowing it, and as you have read through this book, you can think of a time you did something similar to solve a problem.

A combination of these steps should be enough to solve a problem even if it culminates in a workaround.

STEP 11.
THE SECRET STEP

Equipping yourself with experience and now, the knowledge of this book, apply it in a way to maximize your success.

How do you apply what you have learned to your real life, today?

ENGINEERING IN REAL LIFE

Engineering IRL
Experience is king.

There's a reason why the number of years worked in engineering so often gets looked at favourably, and it's because you can't simulate in a learning environment the types of problems you face in the real world. Just because someone has five years of experience, it doesn't indicate how well they did the job.

It is a flawed mentality, but it is reality.

With this book, you will need to use the practical application of the principles learned explicitly for your domain.

You continue to build your lessons learned, not just for a specific project, but also for you, the engineer. Keep these lessons you collate as if they were your assets. They have value, and it grows exponentially over time.

As you complete projects, the artefacts and documentation you produce belong to the company.

You can't take these designs to your next job and re-use it there. In most cases, you can't even take similar products or processes either. You might not be able to go to a competitor for a while, depending on your contract.

So what can you take?

Your experience. Your knowledge and skills. The principles and lessons learned. The technical prowess and your soft skills. Your confidence and leadership. Your communication skills and your networks.

Similar to your degree, no one is taking these away. You may have noticed that every step came with a type of professionalism and approachability, which makes people want to work with you—someone they can trust.

Imagine, by increasing trust in you and your abilities you get assigned to more significant projects. Next, apply this type of thinking to your real-life for success in both your career and overall life.

Practical Application of 10+1

Although I clearly state this is a guide, some might find their specific problem is not solved. And fair enough.

If you had a problem with your car engine right now, this isn't a lookup for engine fixes. Instead, the idea is to apply these principles or check one of these steps.

If you do all of these steps and you're telling me nothing will work, then you did not do Step 1 completely. Return there.

You can work with so many combinations here. Keep practising handling your own problems and also, the art of understanding where a colleague might be up to in their problem-solving process. Each problem is different, so there is no "one-size-fits-all" method. Therefore a guideline with flexibility gives you options on what to try next.

You decide where you should go.

Lessons Learned

Build your database of lessons learned. What you use to store these can be any technology which suits you best.

You get a few things from doing this. At best, you have a continuous log of engineering work, a list of problems solved you can use for further engineering certifications, you can move jobs and take your core achievements to build a portfolio.

You can maintain your skills with practice, and your knowledge you maintain by storing and educating.

At worst, the very exercise of you writing down the lessons learned increases your chances of remembering it because of the formal brain processes which take place, essentially explaining it to your future self and storing it externally.

The process is called encoding, and modern neuroscience tells us as we perceive things, the brain sends the information to the hippocampus for analysis.

It's where a long term memory is decided and writing things down helps with the encoding process.

Plus, you know you understood something if you can teach it.

Quantify Your Pay Rise

It is rare for an engineer to get a pay rise without asking. What is even more unlikely is an engineer laying out the pre-requisites in their mind that allows them to ask for one, as they do not know how to quantify it. They logically disqualify themselves from asking for appropriate compensation.

Engineers are great problem solvers because they are logical. The problem with logic is outputs are dependent on the inputs.

So when you have no inputs to work with, you have no answers. The inputs here are the requirements for asking for a pay rise.

Without this input, engineers don't know what to do and typically just take what comes to them. This lack of input data is because there are no scientific principles, no rules for them to reason.

Here's how I can give you one of those inputs to start. By solving problems better and more consistently you begin to outperform your fellow engineers, if you can show better results, you can use them to lead to a conversation about remuneration.

If you listen thoroughly to 10+1, then you have maintained your lessons learned and your obvious quick fixes and all of a sudden you have a workable log of what you bring to the company and the role. Effectively your value.

It is crucial to set up a discussion before your review – if you wait until your assessment, it is too late.

Typically, you would bring this up the previous year or better yet, during an ad-hoc meeting with your manager to review how you are tracking.

You can have a conversation and ask, what is the process or what would you have to achieve this year to get a pay rise? As you solve problems and document your strategy, while using this knowledge to help colleagues, you can quantify your pay rise.

It varies per company, again, there are no rules, but you can find a correlation with how a company compensates its engineers and your perceived value or your official roles and responsibilities.

Bigger projects typically tend toward more roles and responsibility.

Get on Bigger Projects

Similar to justifying to yourself why you deserve a pay rise, you can also get yourself on more significant projects, gaining more responsibility, which feeds into a more obvious path to getting more pay.

This book gave you the ability to problem-solve better, but also a sequence of questions allowing you to help your colleagues solve their problems too. You become a logical asset in terms of both your ability and how well you work with others.

Bigger projects always require bigger teams, and those people who enjoy working with others have an advantage when project managers and upper management are discussing and negotiating their resource plans.

Because you are equipped with better problem-solving skills (10+1) and knowledge (your collection of lessons learned), when people come to you, they are there for both you and your valuable resources.

It can be quite rewarding to close more extensive and more complicated projects. Pay is important, but it isn't everything.

Applying to Real Life

Every problem you have can look back at this unofficial methodology almost as a checklist. The more experienced you get with particular issues, you may find you skip steps in this list – and it's okay.

Remember, when you are stuck and under pressure, keeping calm and ensuring you have gone through all of the steps in this book will help get you there.

For engineers, it often seems like you can solve the world's most complex problems, but in real life, you are not able to solve basic problems. There are a few reasons; one being you might not be in a problem-solving mood after you have spent all day working on issues.

However, a more prevalent reason is that in real life, you are typically dealing with emotions, opinions and more unexpected circumstances. These are not such measurable inputs, and therefore, no amount of logic can help solve such problems.

But what did we learn in this book?

We learned how we could approach known problems and unknown problems. We change variables, we look outside the box, we create a trend of the data, and it turns out we can solve spuriously occurring issues.

Seemingly random events.

The real problem is in Step 1, your question. You frame a real-life problem as unsolvable due to the fact they are ungrounded in physics and fall in the category of emotions and opinions.

First, is this still true? And second, strip out the emotions and opinions, is there a better question you could be asking?

Look for the underlying problem.

One other thing you can do is accept an opinion as both true and false and then apply logic to both – Schrodinger's statement. You quickly learn the reason of a result doesn't quite matter as much as the process of understanding both lines of argument. Having empathy and showing understanding.

Finally, although the book focused on the immediate problem-solving, if you were paying attention, we also looked at improving as an engineer, taking care of "future you" and putting yourself in a position to better problem solve in the future and becoming an expert engineer.

When you approach situations for yourself or when helping others, think back to this book and see if you can assist in finding the solution and ultimately – find the problem.

You might even help them out further by suggesting they read this book for themselves.

Remember to try the online problem-solving challenges on our website https://www.engineeringinreallife.com and sign up for free membership to ensure you have all of the 10+1 skills. You can also pick up a handy, official Engineering IRL check sheet print off available for you.

Imagine, engineers with the capabilities and confidence to work maximally to their potential while also helping each other solve problems across teams, departments, companies and industries. Large groups working together to solve complex problems can achieve some pretty amazing things.

Working together, we can solve problems across the spectrum problems and dare I say, *any* problems you will face.

And that's pretty powerful.

```
/*******************
 * End of Book Content.
 * Fixed bug which caused
 * infinite loop of steps.
 * A.Sario 2020
 *******************/
```

If this book has given you insights or helped you solve problems in any way, please pass it on to someone you want to help.

CHEAT SHEETS

10+1 Steps Summary

Use the 10+1 Method to come up with one or more solutions.

For the problem you are having, run it through this process and once you have an answer, use it to form your internal database of solutions.

1) The Question
   a) Are we asking the right question?
2) The Obvious
   a) Have we applied the known steps to resolve?
3) Eyes
   a) How do we know our assumptions are true?
4) Check Yourself
   a) Identify what Layer 1 is of your problem
5) Doctor G
   a) Is there a part of the question available online?
6) The RTFM Protocol
   a) Have I read the manual?
7) Strip
   a) Have I removed variables and checked result?
8) What about the Environment?
   a) What are the surrounding variables of my situation?
9) Phone-A-Friend
   a) Are there any other engineers who have dealt with a similar problem?
10) Pray
    a) Have I spoken to a duck?
11) The Secret Step
    a) What Shouldn't Work?

The Question

- ☐ Are we asking the right question?
- ☐ Do we have the root of the question?
- ☐ Have we used a combination of open and closed questions?
- ☐ Did we consider the drivers and perspectives of the people involved?
- ☐ What are our biases, if any?
- ☐ What are the fundamental truths?
- ☐ What are the narratives of the situation and my next step?

The Obvious

- ☐ Have we applied the known steps to resolve?
- ☐ Did we try doing exactly what the clues say?
- ☐ Did we follow our basic procedures in detail?
- ☐ Did we try things to see what sticks?
- ☐ What would an expert check?
- ☐ Is my ego or pride preventing me from performing a quick check?
- ☐ Take note of the "obvious" checks and develop your list of known fixes.

Eyes

- ☐ How do we know our assumptions are accurate?
- ☐ Do we have the correct sensors, indicators, logging, alarms and reports?
- ☐ Can we monitor other pieces of the equation?
- ☐ Have we zoomed out and taken a top-down view of the overall situation?
- ☐ Walk-through the journey of the point of interest when things are working.

☐ Walk-through the journey of the point of interest when things fail.

☐ Use your sixth sense. Utilize different sensors, measurement, tools or monitoring types.

Check Yourself

☐ Identify what Layer 1 is of your problem

☐ Can you restart a system on layer 1?

☐ Check if the device is connected, if you are looking at the right object or device in the correct location.

☐ Have turned it off and on again?

☐ Did we consider replacing the medium? (E.g. cables, power supplies, etc.)

☐ Avoid adding unnecessary variables to your situation.

Doctor G

☐ Is there a part of the question available online?

☐ Have I used the 4-elements in searching?

☐ Are there other forums or online communities that specialize in this field?

☐ Have I gathered the data I can by checking myself?

The RTFM Protocol

☐ Have I read the manual?

☐ Have I checked all supplied documentation?

☐ What tests will give me more clues to the root cause?

☐ Have I made my 8-step investigation plan and answered those questions?

☐ Open the box and check

Strip

- ☐ Have I removed variables and checked result?
- ☐ Do I know for sure 1 = 1?
- ☐ What are the universal truths of this situation?
- ☐ Prepare a list of tests and perform rapid Strip testing
- ☐ Have I reduced the problem to something so simple it works

What about the Environment?

- ☐ What are the surrounding variables of my situation?
- ☐ Has timing, weather, the person, the temperature got any impact?
- ☐ What are the relationships in my situation?
- ☐ Have I got a drawing to model the problem?
- ☐ Did any other projects take place recently that changed the circumstances?

Phone-A-Friend

- ☐ Have I done the problem-solving steps but still need help?
- ☐ Are there any other engineers who have dealt with a similar problem?
- ☐ What response do I need?
- ☐ Have I considered the urgency, traceability and repeatability of my situation?
- ☐ Is my logic sound?
- ☐ Can I attack my syllogism at the proposition?

Pray

- [ ] Have I spoken to a duck?
- [ ] Have I laid out the sequence of events?
- [ ] Can I visualize the problem?
- [ ] Employ your subconscious to help.
- [ ] Have I detached from the problem?
- [ ] Utilize creative thinking instead of focused thinking

The Secret Step

- [ ] Have I tried everything in the ten steps?
- [ ] Am I in the right mindset to solve this?
- [ ] What shouldn't work?
- [ ] What isn't the problem?
- [ ] Can I trigger correlations?
- [ ] Report the situation to the relevant stakeholders
- [ ] Look for the clusters of problems
- [ ] Utilize the Engineering Hierarchy of Controls

Add your own and share it with Engineering IRL.

☐ _____

☐ _____

☐ _____

☐ _____

☐ _____

☐ _____

☐ _____

☐ _____

☐ _____

☐ _____

☐ _____

☐ _____

☐ _____

☐ _____

☐ _____

☐ _____

ACKNOWLEDGEMENTS

My Wife, Maria Pharrah Jhae Acierda Sario for helping me balance everything to get this book done.

Andrew Sparks, Saira Arias, Keith Lam, Taimoor Khan, Robin Haidar for working through the book design, story and editing.

John Acierda and everyone at Lowkey for their hard work on cover design, branding and marketing.

All the engineers I've worked with over the years.

And finally, for everyone who has read this book and passed it on to someone you thought it could help, regardless if they were an engineer or not, thank you.

-Andrew

ACKNOWLEDGEMENTS

INDEX

INDEX

ABOUT THE AUTHOR

Andrew Sario, born on December 3, 1988, in Australia is a Professional Systems Engineer and the creator of the Engineering IRL website, where he makes content and publishes engineering related articles, books videos and courses with member community resources.

His parents, who migrated to Australia from the Philippines in 1987, were very family-oriented and two prominent influences in Andrew's life. With his Father, Rolyn, being a Civil Engineer and his Mother, Erlina, a Business owner in the Philippines setting a pathway subconsciously for his future.

Andrew is a Husband and a Father who owes much of his success to the support of his loving family and wife Maria, who challenges him to reach his potential.

He is a graduate of the University of Technology, Sydney where he received a Bachelor's degree in Computer Systems Engineering with a Diploma in Engineering Practice and has since worked across multiple industries in multiple roles of Engineering.

Primarily working in critical infrastructure and dealing with operational technology, Andrew has gained specializations in control systems, networking, software development, innovation, programmable logic controllers and cybersecurity.

He went through several role changes from intern to lead engineering positions which presented many challenges along the way. This lead to a desire to help others achieve their goals in engineering, too. Andrew has mentored many engineers officially and unofficially culminating in his work with Engineering IRL, in addition to his primary duties as an Engineer.

As an author, he focuses on engineering concepts for a wide range of perspectives, including technical and engineering books for kids through to general engineering philosophies and mindsets.

You can get access to the type of content he produces in both the articles and the podcast content found at engineeringinreallife.com.